20 008 703 17

AF616535

WITHDRAWN

VIBRATION OF BEARINGS

APPLICATIONS OF VIBRATION SERIES

K. Ragulskis, Editor

K. Ragulskis, R. Bansevičius, R. Barauskas, and G. Kulvietis, Vibromotors for Precision Microrobots

P. Alabuzhev, A. Gritchin, L. Kim, G. Migirenko, V. Chon, and P. Stepanov, Vibration Protecting and Measuring Systems with Quasi-Zero Stiffness

S. Korablev, V. Shapin, and Yu. Filatov, Vibration Diagnostics in Precision Instruments

I. Vulfson, Vibroactivity of Branched and Ring Structured Mechanical Drives

K. Ragulskis and Yu. Yurkauskas, Vibration of Bearings

VIBRATION OF BEARINGS

K. M. Ragulskis
A. Yu. Yurkauskas
Kaunas Polytechnic Institute
Kaunas, Lithuanian SSR

English Edition Editor
E. Rivin
Department of Mechanical Engineering
Wayne State University
Detroit, Michigan

HEMISPHERE PUBLISHING CORPORATION
A member of the Taylor & Francis Group
New York Washington Philadelphia London

VIBRATION OF BEARINGS

Originally published as Vibratsiya podshipnikov by Mashinostroyeniye, Leningrad, 1985.

1 2 3 4 5 6 7 8 9 0 BRBR 8 9 8 7 6 5 4 3 2 1 0 9

This book was set in Times Roman by Hemisphere Publishing Corporation. The editor was Victoria Danahy; the production supervisor was Peggy Rote; and the typesetter was Wayne Hutchins.
Cover design by Renee E. Winfield.
Braun-Brumfield was printer and binder.

Library of Congress Cataloging-in-Publication Data

Ragul'skis, K. M. (Kazimeras Mikolovich)
[Vibratsiia podshipnikov. English]
Vibration of bearings / K. M. Ragulskis, A. Yu. Yurkauskas; English edition editor Eugene Rivin.
p. cm. — (Applications of vibration)
Translation of: Vibratsiia podshipnikov.
Bibliography: p.
Includes index.
1. Bearings (Machinery)—Vibration. I. IUrkauskas, A. IU. (Al'girdas IUozovich) II. Rivin, Eugene I. III. Title.
IV. Series.
TJ1061.R3413 1989 88-16581
621.8′22—dc19 CIP

ISBN 0-89116-829-X
ISSN 0897-8301

CONTENTS

PREFACE

The continuous progress of some significant branches in technology (the manufacture of flight vehicles, instruments and machine tools, automotive, electric machine engineering, and other industries) is linked with the use of numerous rolling contact and sliding bearings. In many cases, performance of the instruments and devices is highly dependent on dynamic phenomena taking place in bearings. That is why the questions of technical diagnosis of the whole unit are so inseparably related to the diagnosis of the bearing assembly, which by itself represents an autonomous rotory system which is the basic source of all undesirable disturbances in the machine or instrument as a whole. This can be exemplified by gyroscopic instruments in which the friction torque in the bearings causes appreciable deterioration of the dynamic characteristics of the whole setup due to vibrations. In tape recorders, oscillations in the speed of tape movement measured by the coefficient of detonation are caused by defects in the geometric sizes, accuracy of rotation, and variation in friction torques in the ball bearings used as the rolling contact support of all the rotating components (tone shaft, guide rollers, etc.).

In recent years, extensive studies aimed at determining dynamic characteristics of the bearings and rotor systems working under different conditions have been carried out.

All phenomena taking place in the bearing assemblies are to greater or lesser extent related to vibrations. Consequently, a change in the friction torque, varying parameters of a lubricating film in bearings, oscillations of rings, and other processes may be treated as vibrational changes. Methods of analysis and data processing for these processes are identical and thus all the processes in this book are considered under the common headline of vibratory processes.

To make our diagnostic ability more efficient necessitates first studying the processes which take place in bearings and bearing assemblies.

The tasks posed to the industry may successfully be solved by improving the quality and expanding the capabilities of technical means for the control and measurement of parameters of bearings and rotor systems.

Registration of the parameters, however, has only an auxiliary character, because the basic goal of measurements is to facilitate the diagnosis of the object being tested by taking into account its dynamic characteristics.

Ball bearing supports and bearing assemblies may be diagnosed on the basis of various characteristics. One of the most reliable techniques is the diagnosis of the bearings and their assemblies using the following parameters: friction torque, hydrodynamic oil film, vibratory characteristics, and temperature levels of the interacting components in the bearing assemblies. The interrelation between the state of the lubricating layer and working capacity of the bearing assembly is investigated.

All the investigations being carried out are aimed at determining the technical state of the bearings and their assemblies and ensuring their further safe operation in a given assembly or machine.

This book analyzes the factors affecting the working characteristics of bearings and shows the basic techniques and means of measurement for the processes taking place in bearings. A methodology for an analytical study of bearings characteristics and some results of the experimental investigation dealing with rolling contact bearings are also included in this book.

CHAPTER

ONE

A SURVEY OF METHODS FOR MONITORING AND DIAGNOSIS OF BEARINGS

Control of the processes in rolling-contact bearings is accomplished by diagnosing their technical condition.

Diagnostics means determination of a technical condition of the system under test (in this case, of the bearing assembly) with an aim of determining its performance and the remaining resource of its working life. Since the ultimate goal of diagnostics is to improve reliability and to extend the working life of a bearing assembly or rotor system, the most critical stage of the process is the correct selection of characteristics for the diagnosis, on the basis of which it would be possible to determine the technical condition of the complete assembly.

It is well known that common criteria for all cases of trouble cannot be established. However, selection of the appropriate criteria for each group of mechanisms or systems is feasible. The correct diagnosis of one unit, especially a bearing unit, frequently determines performance of the whole mechanism.

An assembled bearing unit can be checked only by a diagnostic technique not involving disassembly, i.e., in the circumstances when information is extremely difficult to obtain. In such cases, one characteristic is often not enough for the diagnosis of performance and the resource of working life of the assembly, and a combination of several characteristics has to be taken simultaneously using the statistical methods for data processing and extracting of information. The book, therefore, is not only concerned with purely diagnostic problems, but also deals with other related issues without which the solution of the main problem would not be possible.

1.1 CLASSIFICATION OF METHODS

Longevity, reliability, and diagnostics are closely interrelated parameters determining the technical condition of machines and mechanisms as a whole. Diagnostics, being based on a sufficiently developed scientific foundation, on mathematical and physical methods enabling us to obtain optimum results, is a new rapidly developing branch of technical cybernetics. In addition to that, diagnostics has become an integral part of technical maintenance because it allows determination of the condition of machines without their disassembly and facilitates prediction of the remaining resource of their useful life. The parameters of bearings, bearing assemblies, and rotor systems may be evaluated by a variety of methods and checked both by standard and specialized instruments and devices.

The methods for the technical diagnostics of bearing assemblies as well as other mechanical systems are as follows: visual control; static diagnostics; dynamic diagnostics; system of test diagnostics.

Diagnostic means may be classified according to the classification of methods, though we should not forget that some of them have multi-purpose functions. Therefore, the means of diagnostics include: visual control, and measuring, universal, specialized, and automated transducers.

As it has already been mentioned, the methods and instruments under consideration mostly deal with the diagnostic parameters which have been accepted as the main criteria in determining the state of bearing assemblies and separate bearings, i.e., friction torque, vibration, temperature, parameters of an oil film, registration of stress waves (acoustic emission).

The following requirements have to be satisfied by these methods and instruments.

1. High accuracy of measurements exceeding the accuracy of the instruments in which the bearings and rotational systems are built in.
2. High sensitivity that allows measurement of the parameters of miniature precision bearings.
3. Conditions of measurement that must fully represent the operational regimes of the bearings.
4. Minimum inertia of the sensing element of the measuring device.

5. Minimum nonlinearity. The working conditions of many bearings differ so much that friction torque, vibrations, and other parameters during operation sometimes change by more than two orders of magnitude. Hence the linear range of measuring instruments should be large.
6. Universality and simplicity of adjustment for different types of bearings (this is especially important for shop floor control).
7. The possibility of measurements with simultaneous recording of reading.
8. High productivity of measurements.

Such high demands sometimes call for different approaches for each individual case so that the principles and methods of measurements vary for apparently similar measurements.

The study on bearing assemblics and rotational systems for the analysis of their dynamic properties and determination of performance is carried out in two directions:

1. The systems under test are identified, i.e., an approximate mathematical model is constructed and its analysis for obtaining the optimal operating indices is performed.

2. The bearing assemblies and rotational systems are being diagnosed with the purpose of evaluating the quality of manufacture, determining their performance capabilities and their remaining service life, and preventing possible premature failures.

Both directions are based on the measurement of certain actual parameters: vibrations, friction torques, hydrodynamic oil film, etc.

For investigation of rotational systems being characterized by excitations with a wide frequency range, the vibration measurement is taken as a basis. Since vibrations frequently consist of the sum of harmonic and random components,

$$U_{(t)} = \sum_{i=1}^{n} B_i \cos(\omega_i t + \varphi_i) + \xi_i(t)$$

their analysis must be carried out by the methods of statistical data processing.

The methods of visual control. These methods should be considered in two directions. The first direction is a primitive visual inspection and is aimed at immediate discovery of faults. It includes a preliminary control check of the surface finish of rolling elements, their geometric parameters, noise, run out, smoothness of rotation based on engineering experience and intuition. Such methods are good for checking bearing elements with the aid of microscopes and other instruments, for visual inspection and diagnostics using some characteristics already obtained earlier. However, these methods cannot be recommended in the case of an expert check and diagnosis of rotational systems for certification or for compliance with specifications.

In the past few years the ways of direct monitoring of the crucial assemblies (including rotational systems) situated in the positions of hard access have been developed. Use of flexible optical fiber with stroboscopic light sources increases potential of control. It allows the continuous inspection of tightly packaged assemblies with high speed rotational units. In addition, this method of inspection makes it possible to investigate the process of initiating and developing of microscopic fatigue cracks, which are recorded on a photographic paper or film without stopping the machine.

An interesting instrument is worth mentioning [20] which is used for fatigue testing. A conforming package of glued together endfaces of optical fibers is attached to the working segment of the object being tested. Axes of these fibers are arranged normally to the surfaces of micro-areas of the object. The opposite ends of the fibers are spread in one plane and glued together in such a way that their mutual arrangement is the same as in the contacting package. Thus, a rectangular screen is formed onto which a full-scale curved linearized surface of the object under test is being projected. The light reflected from the object is conveyed through the same light guide back to the screen. By changing the phase of the stroboscopic lighting, it is possible to obtain an image of the working segment of the object being tested at any moment of the dynamic loading, for example, at the moments of a maximum opening or closing of fatigue cracks, etc.

In spite of some advantages, methods of visual inspection are not very promising. Those methods which yield good results and are widely used have been extensively described in specialized technical and scientific publications.

Static methods of diagnostics. Up to the present time these methods are the most popular both in industry and to a somewhat lesser extent in research establishments. By static methods it is possible to determine: radial and axial "play" of bearings; misalignment of endfaces of bearing races during their mounting into the bearing assembly; static load carrying capacity of the elements and whole bearings; contact deformation, etc. The static methods can also be used for determining the starting torque.

For static measurements the metallography methods are used, especially for the determination of the qualitative characteristics of the bearing elements. The methods of the measurement of linear quantities are applied when the operational condition of the bearing can be determined by its compliance with the specified geometric parameters. There are also methods of electrical measurement meant for linear and mechanical parameters. Starting torque or static moment of friction is mostly determined by the compensation method. Static measurements of parameters can be represented by the simplest flow chart of the diagnostic system (Fig. 1.1).

Such factors as load (axial or radial), temperature regime, a lubricating or oxide film, etc., are always applied to the input of the system, i.e., the bearing or the bearing assembly.

At the output, using the static methods of diagnostics, usually only one

Figure 1.1 Block diagram of diagnostic system: x_1, x_2, . . . , x_n are the factors of the external action; y_n is the criterial parameter.

parameter is being picked up, which under this method of diagnostics is considered as the criterial one, i.e., $y_n = Ax_n$, where A is the significance factor of the parameter.

From the point of view of the general methods for obtaining measurement results, according to the universally adopted classification [37], the static control methods can be either direct or indirect.

Under direct measurements the unknown quantity is determined directly from the instrument scale, and under indirect measurements this is done from the readings of the direct measurements of one or several parameters having a known correlation, for example, the measurement of the angles by the two sides, by the side and hypotenuse, etc. (i.e., with the help of trigonometric functions).

Each measurement can be carried out both by absolute and relative methods. With an absolute method, a dimension being measured is directly determined from the instrument readings. The relative (comparative) measurement directly indicates only the dimension deviation from the reference measure or a test piece, which was used to set the instrument to zero. In this case the determination of the size is performed by algebraic summation of the reference measure and readings of the instrument during the measurement.

The static measurement methods can also be divided into complex and differential. The complex measurements consist of the comparison of the actual contour of the test object with its limit contours determined by the dimensions and tolerance zones of separate elements of that object. They provide the check for the accumulated errors of the interrelated elements of the object limited by the overall tolerance. The differential measurements are reduced to the independent check of each individual element. The complex measurements are mainly used for checking the final products. The differential measurements are good for checking measuring tools and revealing the causes of the dimensional errors of the products. Each of these methods mentioned can be contact and contactless.

Dynamic methods of diagnostics. Dynamic diagnostics means the establishment of certain dynamic characteristics of bearings and their subassemblies by which the diagnosis of their technical condition can be made or the life can be predicted.

Dynamic diagnostics (regardless of the parameter being measured) involves the statistical data processing with subsequent application of recognition algorithms in accordance with the goals of diagnostics.

The characteristic signals on the basis of which the control of the bearings technical condition can be carried out have a complex structure. Therefore, an adequate signal processing has to be performed for its evaluation. The obtained signal may be represented as follows:

$$S(t) = k(t) S_1(t) + \dot{m}(t)$$

where $S_1(t)$ is the useful part of the signal; $k(t)$ and $m(t)$ is the noise.

The signal from a bearing or a bearing assembly has noise (multiplicative and additive interference) resulting from the unstable external conditions. For example, the fluctuations of speed and load introduce the multiplicative disturbances. The superposition on the signal generated by the unit being tested of the oscillations which are excited by other kinematic links of the mechanism would introduce an additive disturbance. The effects of the waviness of ball races, the oval shape of races, characteristics of bearings mounting on the shaft, etc. should be considered as additive disturbances unless their determination is the object of diagnosis.

In the development of diagnostic methods we are faced with a great variety of problems. The most essential of them are: selection of an optimal system of the signal parameters which is distorted by the disturbances to a lesser degree, and the choice of an optimal method for the signal processing at which the disturbance (noise) level is being minimized.

In the general case the solution of the first problem is reduced to a search for the parameters of the signal, the most sensitive for the changing condition of the mechanism, and the least sensitive to different uncontrollable fluctuations. The techniques of signal processing are directed to suppression of noise and clearing of the signal from distortions caused by disturbances. In general, as it has already been mentioned, they are reduced to the statistical signal processing.

Fluctuations of the variable component of the friction torque in bearings and the nature of its distribution in one realization is of great interest. The character of the distribution of fluctuations of the d.c. component of the friction torque and the values of variable components is evaluated by the formula

$$F(x) = 1/tT(x)$$

where t is the time of observation of the process; $T(x)$ is the time during which the random process $x(t) < x$.

Discrete analog of the formula is

$$F(x) = N(x)/N$$

where N is the number of the points of discretization of the process; $N(x)$ is the number of the points at which the values of the process satisfy the inequality $x(i\ \Delta t) \leq x$.

In the case where it is necessary to determine to which class of distribution laws belongs the distribution law of the process under test, it is more convenient to use the density distribution. For its determination there are calculated the numbers of hits into the intervals determined by the relationship

$$[x_i;\ x_{i+1}],\quad i=1,\ 2,\ \ldots\ k;\quad x_0=x_{min};\quad x_{i+1}=x_{i+c}$$

where $c = (x_{max} - x_{min})/k$; x_{max} is the largest observed value in the process; k is usually chosen of the order of 12.

Evaluation of the function distribution is represented by a bar chart (hystogram).

To approximate the form and parameters of the distribution, various hypotheses are being verified. There are several criteria for such verification. It is expedient to use the agreement criterion k—the Pearson's χ^2 criterion.

During the verification of the hypothesis the following value is calculated

$$\lambda^2 = 1/N \sum_{j=1}^{k} (N_j - N_{pj})^2/p_j$$

where N_j is the number of hits into corresponding intervals determined during the evaluation of the distribution functions; pj are probabilities of the hits into the i-th interval of the random variable $x(t)$, calculated for the hypothetical law of distribution.

As a hypothetical law of distribution, the normal law is frequently accepted

$$F(x) = 1/\sqrt{2\pi}\,\sigma \int_{-\infty}^{x} \mathrm{e}\,[(x-m)^2/(2\sigma^2)]\,dx$$

where m is the mathematical expectation; and σ is the dispersion of the process. They are determined from the bar chart and are equal, respectively, to:

$$m = \sum_{j=0}^{k-1} (x_j + c/2)\, N_{j+1};\quad \sigma = \sum_{j=0}^{k-1} (x_j + c/2 - m)^2\, n_{j+1}$$

The values determining the deviation from the normal law of distribution are the coefficients of asymmetry and excess. From the bar chart they are calculated using the relationships:

$$A_s = 1/\sigma^3 \sum_{j=0}^{k-1} (x_j + c/2 - m)^3 n_{j+1};\quad E_k = 1/\sigma^4 \sum_{j=0}^{k-1} (x_j + c/2 - m)^4 n_{j+1}$$

They are also adopted during the approximation of the series of the distribution (of the bar chart) by the Edgeworth series

$$F(x) = 0.5 + 0.5\varphi(x) - (A_s/6)\,\varphi''(x) + (E_k/2.4)\,\varphi'''(x)$$

Here $\varphi''(x)$ and $\varphi'''(x)$ are the derivatives of the normal probability density of the second and third order, respectively.

The generalization of the obtained characteristics may be performed both by the methods of spectral-correlation analysis and by purely probability theory methods. In each case, however, it is first necessary to clarify the fundamental criteria of the diagnosis, the possible combination of diagnostic criteria, the level of reliability of the obtained information, and the indicative diagnostic features.

1.2 REQUIREMENTS FOR ROLLING CONTACT BEARINGS

Rotational systems may be considered as complex electromechanical assemblies which include both mechanical and electrical elements (for example, the rotor systems of electrical motors). The systems with which we deal here are purely mechanical, thereby, even separate bearings may be regarded as a rotor system, the more so that currently there exist such bearings which by their structure are no less complex than complete units, for example, the three-race bearings.

In rolling contact bearings the sliding contact is substituted for the rolling contact which leads to reduction of friction losses. The design of rolling contact bearings is such, however, that they cannot entirely do without the components which produce the sliding contact. As a result, complex processes take place in rolling bearings introducing the additional irregularities into the bearing performance.

Numerous requirements are applied to bearings employed in supports and guideways of precision instruments. These requirements may be classified into general and specific.

The general requirements related to the development of mechanical engineering and instrument making are reduction of noise and vibration, as well as of the friction torque; more accurate geometric shapes; increases in longevity, reliability, and wear resistance.

The specific requirements include heat stability, antimagnetism, resistance to corrosion, and a number of other specific requirements, particularly for the bearings of spacecraft as well as for other apparatus designed for working in special conditions.

Decrease in friction torque is one of the most outstanding features for many rolling contact bearings, particularly in the instruments where the torque often determines the qualities of the whole system. Experimental studies and practical experience show that the friction torque components and especially its variable component, is changing with the angular position. These changes are of great significance for the bearings which work at low speeds and at limited angles of turning.

It was shown that the basic cause for the appearance and variation of components of friction torque are the errors of the rolling surfaces (ovality of races and balls and other deviations); microasperities on the surfaces, contamination of the bearing by the finest particles of abrasive wear products, etc., and also inconsistency of microhardness of the rolling surfaces. Taking into account all the facts and considerations mentioned above, the friction torque should be regarded as one of the basic parameters characterizing the performance characteristics of the rolling contact bearing.

The accuracy of geometric shapes and dimensions is regulated by special standards. The accuracy class is chosen depending on the required accuracy of rotation of the machinery units and instruments, which is to a great extent determined by the accuracy of the bearings. Control of the geometric accuracy is generally performed for the special purpose bearings. For control of geometric dimensions of ball races and of other elements of precision bearings as well as of their surface roughness, the out-of-roundness measuring instruments have gained wide recognition. These "Talyrond"-type systems have magnification of 10 to 20 thousand times and accuracy up to 0.08 μm.

Longevity of a bearing depends on magnitude and direction of a load, rotational speed, performance coefficient, and other factors. Longevity is usually calculated on the basis of the equivalent load P and dynamic rated load capacity C from the formula

$$L = (C/P)^p \qquad \text{or} \qquad L_n = 10^6 L/(60n)$$

where $p = 3$ for ball bearings; $p = 10/3$ for roller bearings; n is rpm.

The longevity can be considerably extended by manufacturing the components of bearings from high quality metal, which is produced using vacuum or electroslag remelting, as well as by improved surface finish of the working surface and by introducing higher standards on accuracy of geometric dimensions, e.g., by using the superfinish process. At the same time, the optimum regimes of operation must be maintained: such as of loading, rotational frequencies, and lubrication, in order to extend the operational life of bearings.

Reliability is determined by performance of the bearing. Improvement in reliability is achieved by the same means as in the case of longevity. To enhance reliability of a bearing, other requirements such as maintaining the necessary care for storage, conservation with the subsequent de-conservation, washing, and lubrication have to be complied with.

Wear resistance means resistance of the bearing elements to the forces of friction, both of rolling and sliding, which are developing between the components of a bearing. Of all the elements, the case is subjected to the most extensive wear since the balls have a greater hardness. If the rubbing pairs are wrongly selected, the wear of the cages becomes excessive. Certain sections of races and rolling bodies are also subjected to sliding and wear. The quality of materials, their heat treatment, and correctness of geometric shapes by means of accurate machining are the basic factors assuring increase in wear resistance.

Reduction of noise and vibrations of bearings is achieved by means of increasing and achieving accuracy and correct geometric shapes in rolling and sliding bodies.

Some ellipticity and polygonality of the races of a bearing before it is assembled do not enhance its noise level, but when such a bearing is installed into a unit the noise frequently sharply increases.

To reduce noise and vibration, use of rubber housings and buffer rings is very effective. The lubricants have to be of high viscosity, since the thicker the oil film between the interacting members the less the effect of surface roughness. One shouldn't forget, however, that if the viscosity is too high, then supply of the lubricant into the space between the cage and rotating section is reduced, the lubrication becomes nonuniform, and it leads to development of vibrations of the cage and to other negative phenomena. Special requirements take into account specific features of the working conditions of the product and the bearing installed in it.

With these requirements in mind, bearings should be manufactured from high-temperature-resistant steels, such as E1347 or R18 and special lubrication coatings, etc., should be used. Antimagnetism of bearings is obtained by manufacturing bearings from nonmagnetic stainless steels and berylium bronze (for specialized electric motor drives).

The space environment necessitates compliance with a number of specific requirements: low pressure, ionizing radiation, presence of meteorites, absence of gravitational force, presence of substances in an atomic form, and the action of high and fast-changing temperatures.

Pressure outside the Earth's atmosphere is about 10^{-15} Pa, and in the interplanetary space is below 10^{-18} Pa. Accordingly, the temperature level is usually determined by radiation. Any machine in outer space both absorbs and radiates heat, thereby the different mechanisms of the space vehicle have different temperature depending on rates of absorption and radiation of heat. In the vacuum an intense vaporization of lubricating substances takes place. Liquid and plastic lubricants of common use cannot be employed in the space environment. Vaporization of metals also takes place, and it is commonly calculated from the Lengmuir equation

$$v = p/17.14\sqrt{M/T}$$

where v is the speed of evaporation; $p = C_e - L/(RT)$ (where L is the heat of evaporization; R is the gas constant) is the vapor pressure; T is the temperature; M is the molecular mass.

The equation is based on the assumption that all atoms evaporating from the surface leave it forever. As a result of evaporation at low pressures, the surface film is destroyed, which directly affects friction.

It is known that if surfaces are clean enough, then the friction between them reaches an extremely high level since the friction coefficient, under complete dryness of the samples in the vacuum and at comparatively high temperatures (of the order of 800–900 °C) is increasing as much as 10 times. This has been proved by F. Bowden and D. Taybor.

Many different methods for improving the bearings' performance in outer space have been suggested. For example, in order to reduce friction, helium, molybdenum bi-sulphite, and other substances were proposed as lubricants. For rolling bearings, the best results have been achieved by using silicone liquids and silicone lubricants. Lubrication, like the other issues of maintainance of functional properties of bearings under special conditions, requires a special study and a specific judgement in each case.

In addition to these requirements for the quality of bearings, reducing their size and mass is now of primary importance. This is determined by the fact that different systems and mechanisms, especially precision systems, gradually become more and more complex.

Therefore, in the manufacture of bearings for instruments, two additional requirements should be met: 1) the continuous reduction in the sizes of bearings, and 2) improvement of the working characteristics of bearings in order that they become equal to and even better than bearings of standard dimensions which are used for the instruments.

CHAPTER

TWO

DIAGNOSTIC FEATURES OF BEARINGS AND BEARING UNITS

The basic indications of bearing performance which characterize the change of their dynamic parameters and gradually lead to pitting, increased wear of components, breakage, etc., are changes of friction torque, the level of vibration, and character of electrical resistance of the hydrodynamic lubricating film in dynamics.

In a physical sense, ability to perform changes because of violation of contact endurance, wear resistance, and general strength. As a result of deterioration of wear resistance, contact endurance, and general strength, a change of geometrical dimensions of the rolling elements takes place, waviness and flat faces appear on the ball races, the difference in sizes of the rolling bodies becomes more pronounced, and run out of races increases. The gradual change of the condition of bearings in the general case is causing structural changes of physico-mechanical properties of the material of the rolling surfaces, as well as changes of geometrical dimensions of the elements and the quality of rotation. These changes are characterizing the degree of the loss of performance ability of bearing assemblies in the process of change of resistance to contact loads.

Performance ability of bearing assemblies may be evaluated by determining:

1) structural parameters by means of testing hardness and microhardness of the material, its electrographic studies, by analyzing material with the help of eddy current or metallographic analysis;

2) geometric parameters, by measuring bearing clearances and a change of scatter of dimensions of rolling bodies, and their deviations from required geometrical shape;

3) physical parameters by establishing the friction torque, the level of vibrations, the changes in hydrodynamic lubricating film. The physical parameters are assumed to be the basic ones during the determination of performance ability of bearing assemblies without their disassembly.

2.1 FRICTION TORQUE IN ROLLING CONTACT BEARINGS

Rolling friction in bearings. The resistance to relative motion in rolling contact bearings is due to many factors, the basic one being rolling friction. This was long assumed to be the only resistance to motion in rolling contact bearings. After further studies of rolling contact bearings it was established that the contribution of rolling friction in bearings is small, although its effect on wear and operating temperature regime is significant. These factors are especially important for miniature rolling contact bearings operating in the very accurate mechanisms of servo-systems, magnetic tape recorders, and other precision units.

The first experiments on rolling friction were carried out by Leonardo da Vinci in 1508. The concept of the coefficient of external friction arose as a result of these experiments. Leonardo da Vinci concluded that for the same surface smoothness this coefficient is a constant for different bodies and equals 0.25. In early research the rolling friction was treated as a mechanical process, i.e., the interaction of rough surfaces of absolutely rigid solid bodies was considered. The concept of rolling friction as a mechanical process was introduced by I. Delagir and was further developed and studied by A. Parin, L. Euler, D. Lesley and others.

M. Coulomb was the first to introduce the hypothesis of the dynamic nature of external friction. In 1871 he experimentally obtained the classical relationship of the frictional force F due to rolling along a plane.

$$F = fP/r = \mu P$$

Here f is the coefficient of rolling friction, having the dimension of length; P is the normal load; r is the radius of roller; μ is a coefficient similar to the coefficient of sliding friction.

This relationship between frictional force and normal load, known as the

Amonton-Coulomb law, marked the end of a long period in the development of the science of friction.

In 1876 O. Reynolds was the first to propose the hypothesis of relative slippage. According to it, the frictional force due to rolling of a perfectly elastic body along a perfectly elastic base develops as a result of relative slippage between the contacting surfaces due to their deformation. It can be seen from Fig. 2.1 that in the region *AC*, surface layers of the roller are compressed (in a direction along the area of contact) and in the plane these layers are stretched. In the region *CB* of the contact area these deformations take place in the opposite direction, and as a result, the microslippage occurs.

A. S. Akhmatov, D. Taybor, and G. Tomlinson have shown that slippage in the contact zone is not the only source of frictional loss in rolling. Accordingly, the hypothesis of molecular interaction was proposed. G. Tomlinson and D. Taybor showed experimentally that even during the rolling of two cylinders or two balls of identical size made of identical material loss of energy occurs. According to G. Tomlinson, when two bodies approach each other atomic and molecular interactions appear and energy is spent to overcome them, which accounts for the additional energy loss.

Taking into account the forces of molecular interaction during the rolling of a cylinder along a plane, G. Tomlinson determined the friction coefficient as follows:

$$K = (3/4)\sqrt{(\pi/2)\, ef/\sqrt{Pr\theta}}$$

Here P is the load on the cylinder; r is the radius of the cylinder; e is the crystal lattice constant; f is the coefficient of sliding friction; θ is a function of the elastic constants of the materials of the bodies in contact.

G. Tomlinson carried out experimental verification of the coefficient of rolling friction by introducing an additional coefficient λ related to the coefficient of rolling friction k (mm) by the expression $k = \lambda l$, where l is the moment arm of the force applied to the rolling body. The coefficient λ was deter-

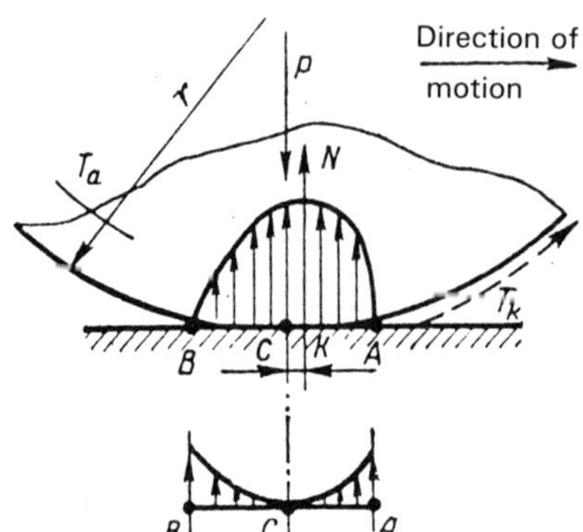

Figure 2.1 Diagram for explanation of O. Reynolds' hypothesis: T_k is the friction torque; T_a is the driving torque; r is the radius of roller; P is the load; N is the normal reaction; K is the coefficient (arm) of rolling friction.

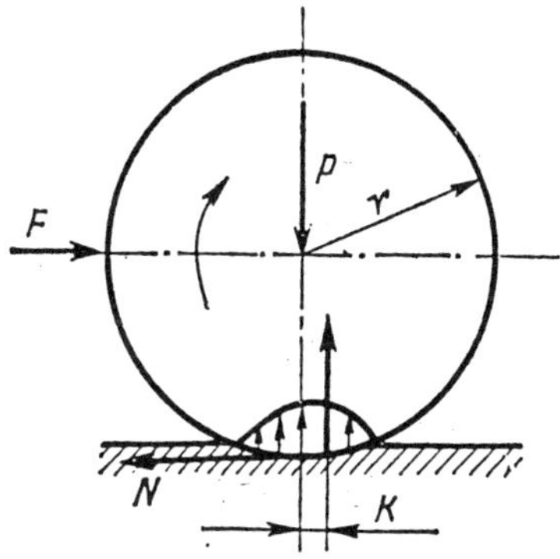

Figure 2.2 Development of frictional force with asymmetric distribution of pressure from roller on ground along contact surface: N is the reaction of a cylinder; P is the load on a cylinder

mined from decaying oscillations of a pendulum. The value of λ thus obtained was twice the theoretical value for amplitudes of pendulum oscillation ranging from 4 to 0.61 ° and this was explained by the fact that the relative displacement and molecular sizes are commensurable.

A. S. Akhmatov developed formulas for determination of the coefficient of rolling friction based on point welding between the friction surfaces. Point welding of metal was established by V. Gardi, F. Bowden and R. Ridler [3].

The works of A. Yu. Ishlinskii, D. Taybor, R. S. Drutkovski, A. Palmgren, and others have great practical importance. They have studied development of internal friction as a result of the imperfect elastic properties of true solids. In order to prove this hypothesis, A. Yu. Ishlinskii considers the rolling resistance in the case of the rolling of a perfectly hard roller along a viscoelastic soil surface, as well as along a soil surface with relaxation. The load distribution on the roller is shown in Fig. 2.2. The presence of friction force in this case is explained by the unsymmetric load distribution.

A. Yu. Ishlinskii developed the relationship for the determination of resistance (at small rpm) in the case of the rolling of a perfectly hard body along a yielding surface:

$$F = P\mu v/(Cr)$$

Here P is the load on the cylinder; v is the rolling rpm; μ is the coefficient of internal friction of the base material; r is the radius of the cylinder; C is the hardness coefficient of the base.

The coefficient of rolling friction is thus obtained as:

$$K = \mu v/C$$

The subsequent development introduced corrections, especially in the mechanical and molecular concepts of external friction. The large amount of work, especially of the applied kind, was devoted to the mechanical concepts. It has enriched our understanding of the laws of elastoplastic deformation and destruction. The latter concept is closely related to the molecular concept, based on the analysis of the physics of the boundary layer and the fine structure of metal at the rubbing surfaces.

Total friction torque in rolling contact bearings. In 1919 A. Palmgren developed a theory of resistance for roller bearings assuming differential slippage as the cause of rolling friction.

The roller bearings widely used now have undergone minor change in form since they first appeared in 1900–1920. In order to determine their characteristics, experimental work has been carried out. The major thrust of research concerned the life, load-bearing capacity, and wear resistance of roller bearings. Research on friction in roller bearings started in 1940 when friction torque became a basic problem and, for certain instrument bearings, the basic criterion determining their usefulness for a particular design.

A. Palmgren's works [33] are among the most important in this field. Load, rpm, type and quantity of lubricant are basic factors affecting the energy loss in rolling contact bearings. According to A. Palmgren, the contact region is a point or a line in the case of a nonloaded bearing. In loaded bearings contact surfaces appear due to elastic deformations. The surfaces are not conjugated either theoretically or from experiments, and hence there appear (small) local displacements and slippage even when the rolling body as a whole is moving as in pure rolling. The energy spent on these displacements is considered lost. The hysteresis loss is added to the losses due to friction arising from elastic slippage. These losses depend on load but not on the approach between the rolling bodies and the ball race.

In 1949 G. V. Poritskii and others extended the theory of friction in roller bearings by analyzing the loss resulting from sliding friction caused by gyroscopic effect. However, Poritskii's studies are too concentrated on this single issue to be considered a general theory of roller bearing friction.

Ito, a Japanese researcher, in 1957 analyzed roller bearing friction under axial load as the friction due to differential slipping. However, he did not consider the friction losses due to slippage arising from the motion of the balls.

In 1959 A. B. Johns assumed motion of the balls of antifriction bearings as the basis for a theory of friction. However, he did not take into account all the factors acting on the roller bearings either, in particular the losses due to elastic hysteresis.

At present there is a tendency to generalize the early theories into a general theory of the overall friction torque of rolling friction bearings. The works of I. A. Spitsin, B. Ya. Kraskovskii, A. I. Sprishevskii, K. Kakuta, R. Barnard, and others encompass all the above-mentioned factors affecting friction torque. Approximate formulas have been developed for analytical determination of friction torques.

When considering friction torque as the result of a number of factors affecting the bearing it is necessary to break down the torque into its components with their subsequent interpretation.

Friction torque in bearings is a complex physical process due to contact and general deformations of contacting bodies, macro- and microgeometry of rolling surfaces, properties of lubricants, resistance of a lubricant layer and the medium surrounding the working elements of a bearing, and physical properties of the materials of the elements in contact.

The friction torque in bearings consists of the following components:

$$T_0 = (T_{sl} + T_{gyr} + T_{hys} + T_{dev} + T_{ca} + T_{lub} + T_{med} + T_{tem})K$$

Here T_{sl} is the friction torque due to differential slippage of rolling bodies in contact areas; T_{gyr} is the friction torque arising from gyroscopic spin or deviation of the axes of rotation of rolling bodies; T_{hys} is the friction torque due to losses on elastic hysteresis in the materials of bodies in contact; T_{dev} is friction torque due to deviations of bearing elements from the true geometric shapes and due to microasperities on the contact surfaces; T_{ca} is the sliding friction torque along the guiding rims orienting the cage and torque arising from the contact of rolling bodies with the cage cavities; T_{lub} is the friction torque due to shear and shifting of the lubricant; T_{med} is the friction torque due to the working medium of the bearing (gas, air, liquid, vacuum); T_{tem} is the complex increase in friction torque due to increase in temperature; K is a correction coefficient taking into account complex change in the friction torque due to the action of forces not taken into account in computing the individual components. These are axial, radial, and rolling forces, vibrational effects, and other unconsidered factors.

Friction torque to differential sliding. Let us consider the friction torque due to local (differential) sliding for the case where the ball rolls along a groove with radius of curvature R in a plane perpendicular to the direction of rolling. Pure rolling will occur along two lines (Fig. 2.3) located on an ellipse of contact at a distance $2a_c$. In other parts of the ellipse there will be sliding between balls and grooves of races because of the unequal distance of contact points from the axes of rotation.

Friction torque due to differential sliding of rolling bodies on the contact area is expressed in terms of work done by the rolling friction bearing in a unit of time as a result of differential sliding friction:

$$A = (F_i l_i + F_o l_o)z \tag{2.1}$$

where F_i and F_o are the rolling friction forces caused by differential sliding during the rolling of the ball along the surfaces of inner and outer races respectively; l_i and l_o are the distances traveled in a unit of time by the point of contact of the ball with inner and outer races, respectively; z is the number of balls.

$$l_i = 2\pi R_{l_i} n_b = (\pi/2) d_b[1 - (d_b^2/D_0^2)\cos^2\alpha]n$$

$$l_o = 2\pi R_{l_o}(n - n_b) = (\pi/2) d_b[1 - (d_b^2/D_0^2)\cos^2\alpha]n$$

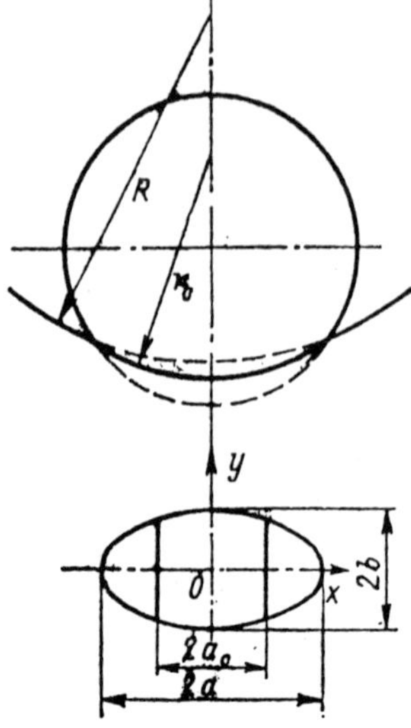

Figure 2.3 Differential sliding: major (*2a*) and minor (*2b*) axes of contact ellipse; r_c is radius of rolling.

The correlation (2.1) is valid for the case where the inner race rotates and the outer one is fixed. In these expressions n is the rotational speed of the inner race of the bearing; n_b is the rotational speed of the ball; D_0 is the pitch circle diameter of the ball; R_{l_i} and R_{l_o} are the radii of curvature of the roller groove of the inner and outer rings, respectively; α is the contact angle.

Then the friction torque resulting from differential sliding of a ball bearing will be

$$T_{sl} = A/(2\pi n)$$

or after expansion

$$T_{sl} = [PD_0 z/(4d_b)][1 - (d_b^2/D_0^2)\cos^2\alpha](r_o B_o - r_i B_i)f$$

Here f is the coefficient of sliding friction; B_i and B_o are coefficients determined by the geometric conditions of deformations; r_i and r_o are the respective radii of deformed surfaces of the ball race of the inner and outer rings.

Friction torque due to gyroscopic spin. Friction torque arising from gyroscopic spin appears when there is a contact angle α on the balls.

To find T_{gyr} it is necessary to determine the moment of inertia of the ball, J from the expression

$$J = (\pi/60)d_b^2(\gamma/g)$$

where γ is the density; for steel, $\gamma = 7.8 \cdot 10^{-4}$ n/m^3.

Friction torque due to gyroscopic spin is

$$T_{gyr} = J\omega_b^2 z \sin\alpha$$

where $\omega_b = \pi n_b/30$ is the angular velocity of the ball rotating around its own axis, and $n_b = n_{sh}(D_0^2 - d_b^2 \cos^2 \alpha)/(2D_0 d_b)$.

Friction torque T_{gyr} increases with an increase in contact angle α and attains a maximum when $\alpha = \pi/2$. Thrust bearings have $\alpha = \pi/2$. In special cases for radial bearings $\alpha = 0$, so that $T_{gyr} = 0$.

In order to avoid gyroscopic spin of balls it is necessary to satisfy the inequality

$$T_{gyr} < Ad_b fz$$

where A is the axial load.

The effect of gyroscopic spin is especially noticeable in radial/thrust bearings.

Friction torque due to elastic hysteresis in the materials of contacting bodies. Energy losses due to elastic hysteresis in the material of the bearing can be determined assuming that during the loading cycle a certain amount of kinetic energy is lost. Friction torque caused by elastic hysteresis has been adequately studied for the case of slow-running rolling friction bearings. However, in the high speed cases, large centrifugal forces arise which are creating additional stresses. Hence in all the expressions developed below, friction torque T_{hys} must be treated as a function of rotational speed.

When the ball rolls along the ball race Hertz's contact ellipse is formed between them (Fig. 2.4).

Energy developed during the rolling of the body about one point of the race is proportional to the load p and to the magnitude of deformation δ.

Contact time t of the rolling body with the race is directly proportional to the dimension of the contact surface in the direction of motion $2b$ and inversely proportional to the circumferential velocity $r\omega$ (Fig. 2.5): $t \approx 2b/(r\omega)$.

Energy loss for the point contact of the rolling body with the race is $T_{hys}\omega t \approx p\delta$.

According to Hertz's theory, where there is point contact between two steel bodies the magnitude of deformation is expressed as

$$\delta = K \sqrt[3]{p^3 \sum \rho}$$

where $\Sigma\rho$ is the sum of the reciprocal of the principal radii of curvature, i.e.,

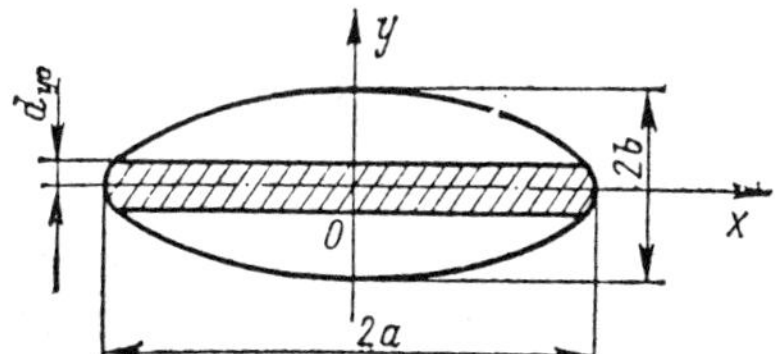

Figure 2.4 Hertz's contact ellipse: major ($2a$) and minor ($2b$) axes of contact ellipse; d_{yo} deformation strip.

$$(\rho_1+\rho_2)_{I}+(\rho_1+\rho_2)_{II}$$

here ρ_1 and ρ_2 are the reciprocals of the principal radii of curvature of the bodies at the point of contact; K is a coefficient depending on the auxiliary functions $F(\rho)$;

$$F(\rho)=\left|[(\rho_1-\rho_2)_I+(\rho_1-\rho_2)_{II}]/\sum\rho\right|$$

Coefficient $F(\rho)$ is usually determined from tables. Approximately,

$$T_{hys} = \Sigma\rho^{4/3} \tag{2.2}$$

The expression (2.2) is valid for point contact when $\delta \approx p^{2/3}$ and $b \approx p^{1/3}$.

A more exact expression for loss caused by elastic hysteresis is given in [21]

$$T_{hys}=1.25\cdot 10^{-4}D_0 d_b^{-2/3}\sum_i p_i^{4/3}$$

where p_i is the load on the i-th ball.

Hysteresis losses are practically independent of the shape of the race.

Friction torque due to geometric errors and the effect of microasperities on contact surfaces. The main cause of changes in friction torque are errors in the geometric shape of the bearing elements (ovality, poligonality of rings and balls, microasperities, etc.).

According to [26] let us write the additional active friction torque:

$$T_{dev} = (dx/dk)N$$

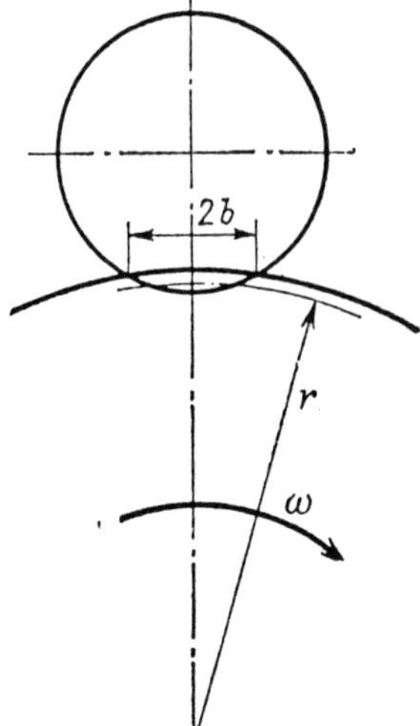

Figure 2.5 Schematic of contact between ball and groove to determine elastic hysteresis losses: *2b*) contact area; ω) angular velocity; *r*) rolling radius.

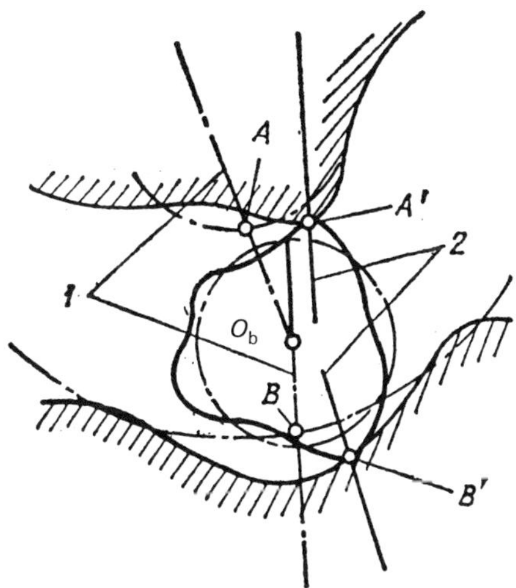

Figure 2.6 Creation of friction torque due to defects in shape of bearing components: *1*, *A*, *B* lines of action and points of tangency in ideal bearing; *2*, *A'*, *B'* in real bearing.

where k is the angular displacement of the moving race of the bearing relative to the stationary one; x is the linear displacement of the moving race of the bearing in the direction of the load N acting on it and caused by deviations from true geometric shape and by microgeometry of the contact surfaces.

The formation process of T_{dev} is shown in Fig. 2.6.

The line of action is formed at every given instant. In an ideal bearing (in which balls and races have ideal shapes), this line *1* passes through the center of the loaded ball. In a real bearing the line of action *2* does not coincide with the direction of the theoretical line of action.

In a real radial bearing, center O (Fig. 2.7) of the race circle of the moving inner ring does not coincide with the theoretical center O_1. Total errors Δz_1 and Δz_2 are caused by geometric inaccuracies. While determining the center O_1, it is assumed that the radial bearing is loaded by purely radial load R applied at the point O.

The radial load is accommodated by two balls which are rolling without slippage. Then

$$T_{dev} = R(dx/dk) = [R/(\sin \alpha\pi/z)]\{(dz_1/dk) \sin (2\pi/z - \alpha_1) - \Delta z_1[d_i/(2D_0)] \cos [(2\pi/z) - \alpha_1] + (dz_2/dk) + \Delta z_2[d_i/(2D_0)] \cos \alpha_1\}$$

where R is the radial load; Δz_1, Δz_2 are the total errors; z is the number of balls; α is the angle between the line of the load application and the center of the ball; D_0 is the diameter of the circle which is drawn through the centers of balls; d_i is the diameter of the race on the inner ring.

Using this approach, it is possible to consider the effect of certain specific errors.

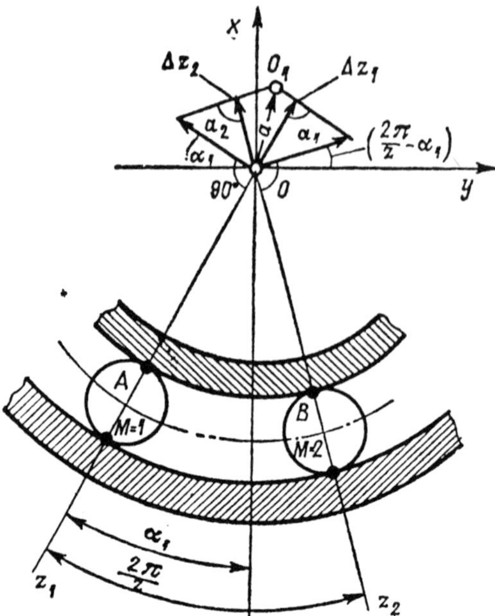

Figure 2.7 Determination of total errors due to geometric defects.

Friction torque due to cage. If we assume that the inner ring of the bearing rotates about a vertical axis and the cage under the action of its own weight is touching the ball at only one point, the friction torque due to the contact of the rolling bodies with the cage cavities (T_{ca}) is expressed as

$$T_{caI} = (D_0/4)[1 - (d_b^2/D_0^2)\cos\alpha]\sin[\alpha + \operatorname{arctg}[d_b \sin\alpha/(2R_d)]G_{ca}f_{ca}$$

where G_{ca} is mass of the cage; R_d is the radius of the race on the inner ring; f_{ca} is the friction coefficient between balls and cage cavities.

The bearing cage is guided by the rim of the inner or outer ring and hence at the time of operation there develops a frictional force as a result of contact of the cage with the guiding rims. Correspondingly, the frictional force is reduced or increased depending on the load pressing the cage to the guiding rims. Depending on which ring (outer or inner) the cage is based (guided) on, the expression is [45]:

$$T_{caII}^{o.r} = 1.38 \times 10^{-4} G_{ca} f_{ca} n^2 D_{o.r} \epsilon [(D_0 - d_b \cos\alpha)/D_0]^2$$

$$T_{caII}^{i.r} = 1.38 \times 10^{-4} G_{ca} f_{ca} n^2 D_{i.r} \epsilon [(D_0 - d_b \cos\alpha)/D_0]^2$$

Here $D_{o.r}$ is the diameter of the outer rim; $D_{i.r}$ is the diameter of the inner rim; ϵ is the eccentricity of the cage with respect to axis of the bearing.

The total T_{ca} is given by

$$T_{ca} = T_{caI} + T_{caII}$$

The housing is in its most desirable position when it is based on the outer ring. In such a case there is an improvement in lubrication in the "flotation" clearance, and thus deformation of the cage due to centrifugal forces is reduced.

Friction torque due to shear and shifting of lubricant. The presence of lubricant between rolling bodies in a bearing leads to additional energy losses due to the viscosity of the lubricant, its physical characteristics, pressure, relative velocity of lubricant flow, temperature regime, and design features of the bearing assembly. P. L. Kapitsa developed the hydrodynamic theory of lubrication at rolling [6].

The layer of oil formed in the bearing prevents direct contact of the rolling bodies, thereby reducing wear and stresses in the metal at points of contact and increasing the contact area.

Power spent on overcoming friction in the intermediate oil layer during rolling per unit length of a rolling friction bearing is

$$W_{\mathrm{M}}^{\mathrm{c}} = 6.5 v^2 \mu / \sqrt{h_0 \alpha}, \quad \alpha l_2^2 \gg h_0$$

where $v = dx/dt$ is the longitudinal velocity of the point of contact during the rolling of the cylinder; μ is the viscosity of the lubricant; h_0 is the clearance filled with lubricant; $\alpha = {}^1/_2(1/R_x \pm 1/R_1)$; here R_x is the radius of curvature of the cylinder surface at point O; R_1 is the radius of curvature of the rolling race.

The cross section of a cylinder with a viscous intermediate layer is shown in Fig. 2.8.

The frictional loss for ball is:

$$W_{\mathrm{M}}^{\mathrm{b}} \approx 6\pi v^2 \mu / \sqrt{\alpha} / [(2\beta - 3\alpha)\sqrt{\beta}] \ln(4m^2\beta)/h_0, \quad 4m^2\beta \gg h_0$$

where 2_m is the thickness of the intermediate oil layer; $\alpha = ½(1/R_x \pm 1/R_1)$; $\beta = ½(1/R_y \pm 1/R_2)$. Here R_y and R_2 are the radii of curvature perpendicular to the direction of rolling.

Power needed to overcome friction in this case is not sensitive to changes

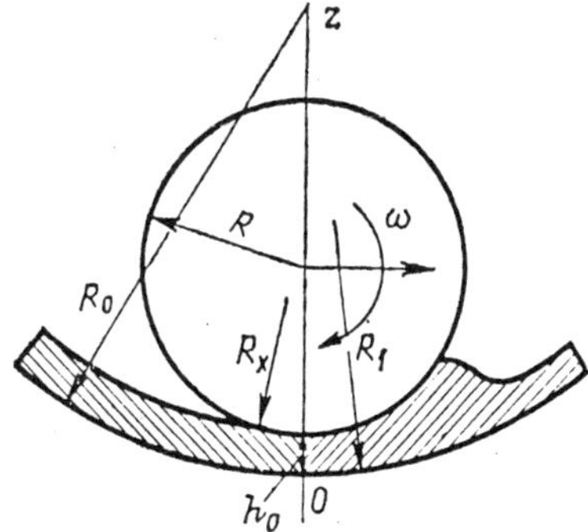

Figure 2.8 Cross section of cylinder with viscous intermediate layer.

in the values of m and h_0 since it has a logarithmic correlation with the thickness of the oil layer.

The resistance torque due to lubricant can be approximately expressed as $T_{lub} \approx z2W_M^b$.

However, it should be remembered that the losses due to friction along the outer and inner rings during the rolling of a ball will be different.

Friction torque caused by the environment of the bearing. The progress of space research made a series of specific demands on rolling friction bearings. In outer space there are many factors that appreciably modify the nature of friction torque: low pressure of the surrounding medium, ionizing radiation, meteorites, absence of gravitation, presence of substances in the atomic state, etc.

It is well known that at low environmental pressure surface films evaporate from rolling bodies. As a result, friction between the clean surfaces reaches high levels due to strong adhesion and the welding between the surfaces. According to experiments of F. Bowden and D. Taybor, when there is a complete dehydration of samples in a vacuum and at high temperatures (800–900 °C) the friction coefficient increases more than ten-fold. To create a protective film under low-pressure conditions it is proposed to utilize helium having low vapor pressure (which can remain in the liquid state in a wide range of temperatures) and good wetting characteristics.

Friction torque caused by temperature increase. Temperature affects friction torque in various ways. With an increase in temperature to 100–120 °C, friction torque decreases, which is explained by the decrease in the viscosity of the lubricant. An increase in temperature beyond 100–140 °C causes an appreciable increase in the constant component of the friction torque as a result of changes in the geometric dimensions of individual components of rolling contact bearings.

The cage material affects the change in the resistance torque with temperature. This is explained by different effects of temperature on various materials. For example, starting from temperatures of 60–70 °C, a cage made of Maslyanite V-2 (porous material soaked in oil) gives out liquid phase which leads to a reduction in frictional torque. An analogous picture is observed in the case of bearings with a cage made of Aman-24 and FN-202 materials.

A more accurate picture for the effects of temperature on friction torque has not been developed.

Analytical determination of total friction torque in rolling contact bearings. So far we have considered the existing theories of friction torque in rolling friction bearings and their components. However, due to their complexities, they are not useful for practical applications, nor do they take into account real life factors that result in development of the friction torque.

For not very important bearing units, approximate expressions for the friction torque are adequate. For this purpose various empirical formulas have been developed.

Let us consider some of the existing expressions, since each gives a more accurate value for friction torque for specific conditions. The suggested expressions should not be used for critical units and mechanisms where, considering the complex nature of the development process for the friction torque, it is necessary to make experimental verification.

The majority of the expressions contain a constant friction coefficient. Although actually it varies over large intervals depending on the working conditions, for approximate computations it is sufficient to use an average value chosen for a particular load and rotational speed n.

The friction coefficient f_F for self-aligning ball bearings is 0.0010; for cylindrical roller bearings 0.0011; for thrust ball bearings 0.0013; for deep groove ball bearings 0.0015; for bearings with tapered and crowned rollers 0.0018; for needle roller bearings 0.0045. These coefficients are used by SKF (Sweden) for computing the friction torque of bearings having an operational life of 10^9 revolutions. Then

$$T = f_F F\,(d/2)\,10^{-2}$$

where F is the load on the bearing, N; d is the bore diameter, m.

According to the laws established by G. Amonton and confirmed by M. Coulomb, the friction coefficient is determined as a ratio of pulling force R to the normal load N: $f = R/N$.

The friction coefficient is found empirically, using friction torque in a bearing instead of pulling force. Friction torque of rolling contact bearings which are part of the bearing units, and friction torque for self-contained bearings, are determined in different ways by various researchers. According to A. Palmgren, the resistance torque is

$$T = T_1(F) + T_0(n)$$

where $$T_1(F) = f_1 d_b (F/C_{st})^c; \qquad T_0(n) = f_0 d_b^2 (\gamma n)^{2/3}$$

Here f_0, f_1 are coefficients depending on bearing design, load, and lubrication method; c is exponent depending on the type of bearing; γ is the kinematic viscosity; C_{st} is the static load carrying capacity of the bearing; d_b is the average ball diameter; z is the number of rolling bodies; F is the load on the bearing; n is the speed of rotation, s^{-1}.

Other researchers use different formulas for the resistance torque but the Palmgren formula is considered to be the best. A. Figatner analyzed experimental results using the Palmgren formula and obtained

$$T = mF + T_s$$

where m is a coefficient; T_s is the resistance torque component not depending on load. The values of the coefficient m by Palmgren and from experimental data (for $n = 170\ s^{-1}$) are shown in Table 2.1.

Considering the large scatter (see Table 2.1) of values of friction coefficients in rolling contact bearings, some researchers either avoid using it altogether in the formulas for determination of friction torque, or take a value based on their own experimental data.

The resistance torque in a bearing for radial loading can be approximately calculated using the empirical formula

$$T = Qfd/2$$

where Q is the radial load on one bearing; f is an effective friction coefficient, equal to 0.004–0.010 (depending on size, type, and operating conditions of the bearing); d is the bore diameter of the bearing.

The expression for the determination of friction torque with the rolling friction coefficient is

$$T = 1.4Q[(D_0/d_b) + 1]\delta$$

where D_0 is the diameter of the race on the inner ring; δ is the friction coefficient, equal to $5\text{–}10 \cdot 10^{-6}$ m.

A more general expression for the approximate determination of friction torque is given in [45]. It should be considered that the coefficient of rolling friction in this expression lowers the actual friction torque.

For pure radial loading $T = T_0 + 1.25\delta(D_0/d_b)R$, where $T_0 = 0.04D_0$.

For a bearing under only axial loads $T = T_0 + 1.5\delta(D_0/d_b)A$.

When both radial and axial loads act simultaneously on a rolling contact bearing, then

$$T = T_0 + (1.5A + 1.25R)\delta D_0/d_b \tag{2.3}$$

All these expressions give an approximate value of friction torque since they do not take into account many factors which are present during the operation of a rolling contact bearing. I. A. Nikitina has developed a formula (based on experimental data) for estimation of the resistance torque in a bearing taking into account lubricants, rpm, and other factors [5]

TABLE 2.1

Radial double row spherical radial ball bearings	2211	2311	1311
Coefficient according to A. Palmgren	0.031	0.025	0.028
Coefficient obtained from experimental data	0.023	0.024	0.020

$$T = -\mu\pi z (D_0/2)\, l_F\, [D_1 n/(30 d_b)]$$
$$\times \{ [4.5 d_b a_0 (d_b h_0 + a_0^2)/(2h_0)]/(a_0^2 + h_0 d_b)$$
$$+ [9 d_b a_0 (d_b h_0 + a_0^2)/(2h_0)]/(4a^2 + h_0 d_b)$$
$$+ [15 (d_b/2)^{3/2}]/(2\sqrt{2h_0} - 9a_0^2 \sqrt{d_b/2})/(2h_0\sqrt{2h_0})$$
$$\times [\tan^{-1} (a_0/\sqrt{h_0 d_b}) + \tan^{-1} (2a_0/\sqrt{h_0 d_b})] + 16.5a_0\}$$

Here μ is the oil viscosity; n is the rpm of the inner ring; D_1 is the diameter of the race on the outer ring; l_F is a coefficient, depending on the curvature radius of the race; h_0 is the minimum thickness of the hydrodynamic oil film; a_0 is the length of the hydrodynamic oil film.

However, this formula is not useful for practical applications because it is complicated and because it contains parameters whose determination in actual conditions is not possible (it is possible to compute these parameters only theoretically). Parameters a_0 and h_0 can be found using the hydrodynamic theory of lubrication developed by P. L. Kapitsa.

It is possible to use the following correlation as a working formula:

$$T = 0.452\, [(1 - \xi)(1 + \xi)^2/\xi]\, D_0^2\, [0.625 \cdot 10^{-3}\sigma/(\lambda_k \sqrt{r})]^{6.5 d_b^{-0.65}} z$$

where $\xi = d_b/D_0$; $\lambda_k = 1/(1 + \xi \sin \beta)$; β is the contact angle; D_0 is diameter of the circle of ball centers, mm; σ is the contact stress, usually taken in the range 85–280 kg/mm^2; r is the cross sectional radius of the race groove.

This formula is recommended for radial-thrust ball bearings working at 50–3,000 rpm.

In modern mechanisms and units subjected to vibrations it is important to know the resistance torque when the bearing functions in vibratory conditions. In this case if the bearing axis is horizontal, the friction torque can be computed as

$$T = \mu_b Q (d_{sh}/2)$$

where Q is the radial load on the bearing; d_{sh} is the journal diameter of the shaft;

$$\mu_b = k[2.8(1 + \alpha)/(d_{sh} g)](D_0/d) V_{av} V_{d.av}$$

where $V_{d.av}$ is the average relative frequency of compacting between the inner ring and balls and the outer ring of the bearing; V_{av} is the average frequency of impacts of the inner ring and balls against the outer ring; k is the rolling friction coefficient; d is the ball diameter; g is the acceleration of gravity; α is the velocity restitution coefficient during impact.

In the case of vertical vibration of a bearing with a vertical rotational axis friction torque is

$$T = f_A Q d_{sh}/2$$

where $f_A = [(D_i/d) + \cos\beta][2k(1+\alpha)]/[f_N d_{sh} g \sin\beta/(V_{d.av} V_{av})]$; D_i is the diameter of the race on the inner ring of the bearing; f_N is the coefficient of nonuniformity of load distribution between balls, usually assumed $f_N = 0.9$; β is an angle at which the load acts on the ball.

According to the data from SKF, the friction torque at the point of contact is

$$m_K/(Qd_m) = (2 - \lambda) \sum_{v=0,1,2} \varphi_{a\gamma} [(\lambda Q)/D^2]^{\gamma/2}$$

where m_K is the friction torque; Q is the load; $\lambda = d/d_m$; d is the diameter of the ball race; d_m is the diameter of the circle passing through the centers of the balls; γ is an ordinal number; $\varphi_{a\gamma}$ is a coefficient; D is the ball diameter.

Coefficient $\varphi_{a\gamma}$ varies over wide limits and depends on the modulus of curvature, elastic properties of a bearing material, shape of the contact surface, and also on set of coefficients. As we see, each specific bearing requires a separate computation which can be performed on the computer.

All these formulas give only an approximate average value of friction torque or contain parameters which can be determined only experimentally.

2.2 HYDRODYNAMICS OF LUBRICATING OIL FILM IN BEARINGS

In considering dynamics of rolling-contact bearings it is necessary to consider the effect of a lubricating layer. The friction torque when a ball rolls along the race is expressed in the general form by the formula

$$T = W/\omega$$

where ω is the angular velocity of the rolling ball; W is the power required for rolling of the ball.

The following formula based on [6] is obtained for power consumption during the rolling of the ball along the race, considering the oil film:

$$W = 17.3\pi v^2 \mu_0 \sqrt{\alpha}/[(2\beta + 3\alpha)\sqrt{\beta}]$$

Here the speed v is practically constant, the viscosity μ_0 can be considered independent of pressure, and α and β are parameters depending on the radii of curvature of the balls and the races:

$$\alpha = (1/r) \pm (1/R_{i,o}), \qquad \beta - (1/r) + (1/r_z)$$

For bodies and ball races of ideal geometric form, the radii r, $R_{i,o}$, and r_z are constants. In actual bearings they are variable. The geometric errors of races can be expressed in terms of a Fourier series. Correspondingly, in the case of the race of the outer ring the error function is

$$\Phi_1(\varphi) = \sum_{k=1}^{m} A_k \cos(k\varphi + \varphi_k)$$

and for the race of the inner ring

$$\Phi_2(\psi) = \sum_{k=1}^{m} B_k \cos(k\psi + \psi_k)$$

where m is the number of harmonics; A_k, B_k are the amplitudes of the harmonic components; φ_k, ψ_k are the corresponding phase angles.

The amplitude of the harmonic components are

$$A_k = \sqrt{a_k^2 + b_k^2}$$

where

$$a_k = (1/\pi) \int_0^{2\pi} \Phi_1 \cos k\varphi \, dx = (1/\pi) \sum_{i=1}^{N} \rho_{H_i} \cos ki \, \Delta\varphi \, \Delta x$$

$$b_k = (1/\pi) \int_0^{2\pi} \Phi_1 \sin k\varphi \, dx = (1/\pi) \sum_{i=1}^{N} \rho_{H_i} \sin ki \, \Delta\varphi \, \Delta x$$

Here $\Delta\varphi$ is the incremental angle (division) of the circular diagram of the race errors, usually $\Delta\varphi = 1°$. Phase angles are expressed as $\varphi_k = \tan^{-1}(b_k/a_k)$.

It follows that the curvature radius varies along the length of the race. The radius of curvature at a given point is obtained by mathematical analysis:

$$R_0(\varphi) = \left\{ R_0 + \sum_{k=1}^{m} A_k \cos(k\varphi + \varphi_k) + \left[\sum_{k=1}^{m} kA_k \sin(k\varphi + \varphi_k) \right]^2 \right\}^{3/2} \Bigg/ R_0 + \sum_{k=1}^{m} A_k \cos(k\varphi + \varphi_k)$$
$$+ 2\left[\sum_{k=1}^{m} kA_k \sin(k\varphi + \varphi_k) \right]^2$$
$$+ \left[R_0 + \sum_{k=1}^{m} A_k \cos(k\varphi + \varphi_k) \right] \left[\sum_{k=1}^{m} k^2 \cos(k\varphi + \varphi_k) \right]$$

An analogous expression is obtained for the race of the inner ring. With a change in the radius of curvature, the geometric coefficients α and β also vary, and consequently the power losses.

However, calculations of the variable component of friction torque (VFT) according to analytical expressions show that the magnitude of this component is negligibly small compared to VFT obtained experimentally. Thus it is not a correct way to describe VFT.

The second approach to the description of VFT is as follows:

Let us assume that the ball rolls between two surfaces (Fig. 2.9*a*). The

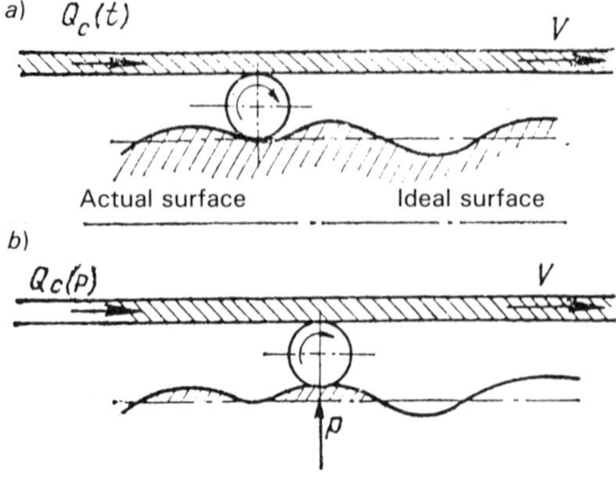

Figure 2.9 Motion of ball along a wavy surface: $Q_c(t)$ is friction torque to rolling without action of force P; $Q_c(P)$ is the same but under action of force P.

lower surface differs slightly from a plane surface and is stationary, the upper surface is an ideal, plane surface and moves with velocity v in such a way that at any moment of time any point on it does not change its position along the height. For the displacement of the upper plane it is necessary to overcome a certain rolling friction force Q_c. The ball will be alternately in the valley or on the peak. During the rolling of the ball across the peak it is necessary to perform a certain work proportional to the magnitude of deformation. The higher the height of peaks, the greater is the amount of work. It is clear that the force Q_c will also change. Here it is assumed that there is no sliding. If the magnitude of a peak depends on its position along the length, the force Q_c will be a function of time. Further, this model helps explain the effect of lubricating oils of different viscosities on $Q_c(t)$. It is well known that the lubricant forms an oily film, thereby increasing the contact area. It is also known that with the increase in viscosity the dimensions of the oil film increase. Consequently, with an increase in viscosity the amount of work done increases, since the deformation takes place in a larger volume. The force Q_c changes proportionally to the work done. So with the increase in the viscosity of the oil the amplitudes of the variable component of friction torque are increasing, while the high-frequency components of the variable resistance torque are eliminated. These assumptions tally very well with the results of experiments, showing that with the increase in viscosity high-frequency components of VFT are damped out but the amplitudes of lower components are increased. However, no expressions are available which describe the dimensions of the intermediate layer of oil because of the difficulty of obtaining them. Consequently it does not seem possible to use this model (Fig. 2.9).

In the model shown in Fig. 2.9*b* it may be seen that $Q_c(t)$ can be compensated by the force developed by the deformation of the peak and acting in the normal direction. The magnitude of the force when the ball approaches the uneven surface depends on the height of the bump and is described by Hertz's theory of contact. It can be written that $Q_c = F(P)$.

Let the friction torque when the ball rolls along the race be expressed by

$$T = \{11.24\mu_0/h_0 a_0^5 b_0 [(r_i + 0.5d_b)/(r_i d_b)]^2$$
$$\times [e^{n\sigma_0}/(n\sigma_0)]^{3/2} [(y/d_0)^4 - 1/(n\sigma_0)(y/a_0)^2 + 3/4\,(n\sigma_0)^{3/2}]\}\,\omega \qquad (2.4)$$

where μ_0 is the dynamic viscosity of the lubricant; h_0 is the thickness of the lubricating layer; a_0, b_0 are the major and minor axes of the Hertz contact ellipse, respectively; r_i, d_b are the radius of the race and diameter of the ball, respectively; n is the piezocoefficient of the lubricant; σ_0 is the maximum stress at the center of the contact area (according to Hertz); ω is the angular velocity of rolling of the ball.

Expression (2.4) has been obtained from the solution of the contact hydrodynamic problem. Since the major axis of the contact ellipse is larger than the minor axis by one order of magnitude, it is assumed that the lubricant flows only in the direction of rolling (plane contact hydrodynamic problem). The expression given for friction torque has been obtained on the basis of resistance to differential slippage when the ball rolls along the race. In this expression the hydrodynamic flow of lubricant, deformation of surfaces, and exponential dependence of viscosity on pressure have been assumed. Formula (2.4) is preferred for the case of heavily loaded surfaces.

In the work [18] a nomogram is suggested to solve the contact hydrodynamic problem of determining the thickness of the lubricating layer between surfaces which are rolling about each other with slippage. The thickness h_0 of the lubricating layer is a function of: total rolling velocity $U_a + U_b$; contact conditions of the surfaces; effective curvature of the profiles $\Sigma\rho$; viscosity of lubricant μ_0 and its piezocoefficient n; and per unit length K_0. It can be directly determined by computing a set of interrelated auxiliary parameters with the help of such nomogram. The accuracy of determining the thickness of lubricating layer h_0 is insufficient when obtained with the smaller scale nomogram but it is very difficult to reproduce the nomogram at a bigger scale. Besides, it is necessary to compute the nondimensional parameters via the respective physical quantities.

Therefore, in order to simplify the computations and to obtain more accurate analysis of dependence of h_0 on a number of basic factors, the following formula is used

$$h_{0\,i(o)} = \{3.17\,[\mu_0(U_a + U_b)]^{0.75} n^{0.6}\}/[K_0^{0.15}(\sum\rho)^{0.45}]$$

where

$$K_0 = [3\sqrt[3]{3}/(8\mu)]\sqrt[3]{P^2E(\sum\rho)^2}; \quad \sum\rho = 4/d_b - 1/R_i \pm 1/R_{i(o)}$$
$$R_i = (0.5D_i + r_i)/\cos\beta_i$$
$$R_o = [(0.5D_i - r_i)/\cos\beta_o] - (r_i - 0.5d_b)(\cos\beta_i/\cos\beta_o) + 0.5d_b$$
$$U_a \approx U_b \approx (\omega_i - \omega_0)D_{ik}/2$$

Here U_a, U_b are the relative speeds of surfaces with respect to the contact point; P is the normal load at the contact point; E is the Young's modulus; ω_i is the angular velocity of the inner ring; ω_0 is the angular velocity of the center of the ball. Subscripts i and o represent inner and outer rings.

To make it easier to determine friction torque at different loads, after a series of transformations expression (2.4) can be written as

$$T = \{6.11\mu_0\bar{s}^6\mu^5\nu/(\bar{k}\bar{t}^{3/2})[(r_i + 0.5d_b)/r_i d_b)]$$
$$\times P^{1.6}e^{3/2\bar{t}P^{1/3}}[(y_{av}/a_0)^4 - (2/3)1/(\bar{t}P^{1/3})(y_{av}/d_0)^2$$
$$+ 0.407 \cdot 1/(\bar{t}^{3/2}P^{1/2})]\}\omega$$

where

$$\bar{t} = n/(\pi s^2\mu\nu); \quad \bar{s} = 1.08 \cdot 10^{-3}1/\sum\rho^{1/3}$$
$$\bar{k} = 3.48\mu^{0.15}\mu_0^{0.75}(U_a + U_b)^{0.75}n^{0.6}/[E^{0.05}(\sum\rho)^{0.45}]$$

μ, υ are coefficients which are functions of elliptical integrals; y_{av}/a_0 is the relative position of the trace of pure rolling; a_0 is determined using the graph or formula which are given in [15].

Computations show that the sum of two terms $(y_{av}/a_0)^4 - (2/3)1/(\bar{t}P^{1/3})(y_{av}/a_0)^2$ does not exceed 5% of the term $0.407(1/\bar{t}^{3/2}P^{1/2})$. Hence we will neglect it in further discussions.

In Fig. 2.10 the dependence of friction torque is shown per unit force for different oil viscosities. From this it can be seen that friction torque even at small sliding velocities sharply increases and can reach astronomical values, which contradicts experiments. This is due to the strong dependence of viscosity on pressure ($\mu = \mu_0 e^{n\sigma_0}$). The dependence of $e^{3/2t}$ on the viscosity of the lubricating oil is shown in Fig. 2.11. It can be seen that the variation in $e^{3/2t}$ is similar to the variation in friction torque. Consequently the term $e^{3/2t}$, which depends on the piezocoefficient and Hertzian pressure, has an important role in the expression for friction torque. Analysis shows that a 20% error in the piezocoefficient of the lubricant alters the above-mentioned term several hundred-fold.

From what has been said it follows that the expression developed in [15] is not applicable in this particular case. In [28] a formula has been obtained for shear stress which takes into account the exponential dependence of the viscosity of the lubricant on pressure and the polytropic dependence on temperature,

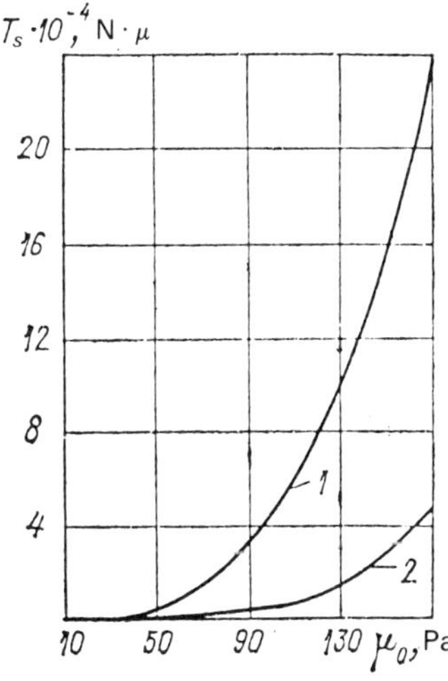

Figure 2.10 Relation between friction torque and viscosity of lubricant under a unit load: *1*) computation; *2*) experiment.

$$\tau = \{(2\sqrt{427\lambda T_s \mu_s})/[h_0 \sqrt{\gamma}\sqrt{k^2\gamma + 2 - 1/(a_1+1)}]\} \times \ln\{\sqrt{a_1+1}\,[k^2\gamma + 1 + \sqrt{k^2\gamma\,[k^2\gamma + 2 - 1/(a_1+1)]}]/\sqrt{k^2\gamma + (1+a_1)}\}$$

Here $k = [(U_a - U_b)\sqrt{\mu_s}]/(2\sqrt{427\lambda T_s})$; T_s is the temperature of the lubricant at the surface, °C; μ_s is the viscosity of the lubricant at the given pressure and temperature T_s; λ is the coefficient of heat conductivity of the lubricant; γ and a_1 are empirical coefficients.

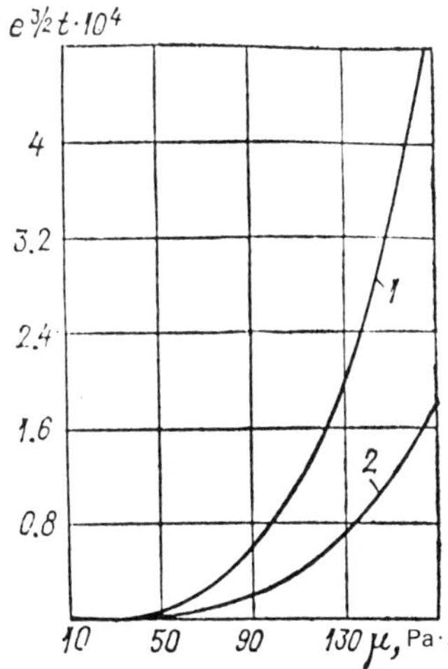

Figure 2.11 Relation between $\mathbf{e}^{(3/2)t}$ and lubricant viscosity under a unit load: *1*) computation; *2*) experiment.

The resistance torque to rolling is

$$T_{\kappa} = \iint_F \tau \, dF \tag{2.5}$$

Formulas (2.4) and (2.5) are difficult to use due to their complexity. In [6, 28] it has been shown that at low sliding speeds and small loads the shear stress can be computed using the classical formula for the isothermal regime $\tau = \mu_0(v/h_0)$.

Let us introduce the following simplifying assumptions:

1) The contact radius of curvature ρ_{av} (Fig. 2.12) is determined approximately by the formula $\rho_{av} = r_i d_b/(2r_i + d_b)$;

2) The instantaneous rolling axis ω of the ball relative to the race passes through the trace of "pure" rolling (rolling without slippage) in the plane passing through the axis of the ring and the center of the ball;

3) Since stresses in the contact zone are relatively small it is possible to assume that viscosity does not depend on pressure [6];

4) The area of contact in the circumferential direction along the axis is assumed to be planar.

According to Fig. 2.12 the distance from any point on the profile of the contact area to the axis of pure rolling ω is

$$z \approx (1/2\rho_{av})y^2$$

The sliding speed at the contact during rolling is

$$v = [1/(2\rho_{av})](y_{av}^2 - y^2)\omega$$

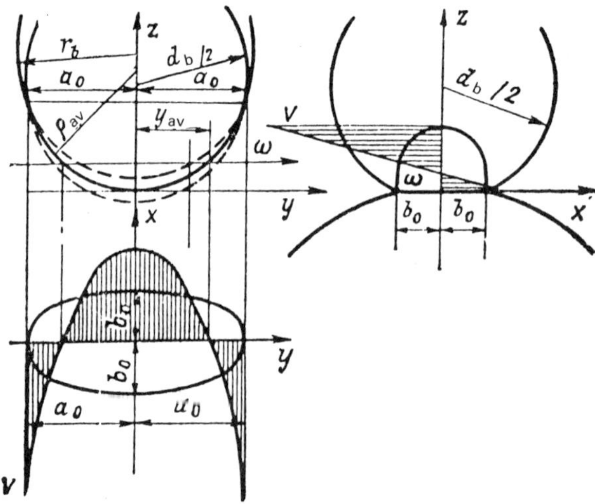

Figure 2.12 Schematic of ball rolling along its race for determination of friction torque due to differential sliding.

Elemental resistance to sliding is

$$dS = \mu_0(v/h_0)\, dx\, dy = \{[\mu_0(y_{av}^2 - y^2)]/(2\rho_{av} h_0)\}\, dx\, dy$$

Elemental friction torque in rolling with slippage

$$dT = dS(z_{av} - z) = (\mu_0/h_0)\omega/(4\rho_{av}^2)(y_{av}^2 - y)\, dx\, dy$$

The friction torque

$$T = 4\int_0^{a_0}\int_0^{b_0} dT = [\mu_0 a_0^5 b_0/(h_0 \rho_{av}^2)]\,[\bar{y}_{av}^4 - (2/3)\,\bar{y}_{av}^2 + 1/5]\,\omega$$

where $\bar{y}_{av} = \bar{y}_{av}/a_0$, i.e., position of the trace of "pure" rolling.

Calculations show that $(\bar{y}_{av} - (2/3)\bar{y}_{av})$ does not exceed 5% of the expression within brackets. Taking this into account, after transformation we get

$$T = [0.8\mu_0 \omega s^6 u^5 v/k][(r_b + 0.5d_b)/(r_b d_b)]^2 P^{2.1} \qquad (2.6)$$

We will express (2.6) as $T = CP^{2.1}$, where C is a constant for the given type and geometric dimensions of bearings.

Power consumed for rotation of the inner ring of the bearing equals $N = N_i + N_o$, where N_i and N_o is power consumed during rotation of the ball along the races of inner and outer rings, respectively.

In order to overcome the resistance, it is necessary to apply a torque to inner ring of the bearing

$$T = T_i + (T_i + T_o)(\omega_b/\omega)$$

where ω_b and ω are angular velocities of the ball and the inner ring of the bearing, respectively.

On the basis of the equation (2.6) for one ball it is possible to write

$$T = [C_i + (C_i + C_o)(\omega_b/\omega)]P_i^{2.1}$$

for bearing with s balls

$$T = [C_i + (C_i + C_o)(\omega_b/\omega)]\sum_{i=1}^{s} P_i^{2.1}$$

Here C_i and C_o are constants for the inner and outer rings, respectively.

In order to separate the variable component in the last expression, it is necessary to relate resistance to geometrical errors in races of rings.

2.3 RELATIVE AND ABSOLUTE VIBRATIONS OF BEARINGS

Vibrations should be considered in two aspects. There are absolute vibrations, that is, vibrations of the complete unit or instrument, and relative vibrations, i.e., vibrations between parts of a bearing, for example, vibrations of outer ring with respect to the inner one. They are able to excite vibrations in other components or units directly attached to the vibrating part of a bearing.

The basic source of vibrations in bearings is a deviation from ideal geometric shape of working surfaces of bearing rings and rolling bodies. It has been established that the effect of active elements of bearings on vibrations is very significant. If an inner ring rotates, then accuracy of its geometric shape and that of the rolling bodies are critically important. According to a traditional classification, vibrations of bearing rings are divided into three groups.

Vibrations with a great wave length (up to 10 oscillations per 1 revolution of a ring) fall under the category of the *first group*. These vibrations arise due to radial runout of rings, which can be as large as 15 μm. In the case of assembly eccentricity, vibration frequency will correspond to the frequency of revolutions.

Vibrations with a medium wavelength (ranging from 10 to 60 oscillations per 1 revolution of a ring) are in the *second group*. The source of these oscillations is the waviness of races. The amplitudes do not exceed 1 μm.

Vibrations with a small wavelength (more than 60 oscillations per 1 revolution of a ring) belong to the *third group*. These oscillations develop due to microasperities on races with the amplitude up to 0.1 μm.

Vibrations of the outer ring are basically due to cyclic variations of loading on a bearing. In this case vibrations arise even in bearings with the most perfect geometric shape. Vibrations appear as a result of rolling along the imperfect surface of a bearing.

Additional sources of vibrations, having a secondary importance, include contacting between the balls and the edge of a race, insufficient height of guiding groove, presence of hard contaminants, slight misalignment, and friction of safety washers or seals against other elements.

Vibrations determined by a variable mechanical compliance are of two kinds:

1. Vibrations with a variable contact compliance when there is a radial load component acting on a thrust axial bearing due to the fact that the elastic shift of the Hertz contacts under the load varies depending on the positive change of a set of balls with respect to the line of the load action.

2. Vibrations due to elastic contact deformation which are significant only in the range of frequencies which are close to the rotational frequency of a bearing, and under heavy radial loads.

Thus, with the increase in waviness of races on outer rings the vibration level of bearings will increase. For example, with a reduction of waviness of races of rings from 2.0 to 0.05 μm, the vibration level decreased by 14 db. Waviness of inner rings exerts greater influence on the level of vibration. Thus, with the reduction of various inner rings from 2.2 to 0.05 μm, the level of vibrations decreased by 20 db. The same is true for waviness of balls. A reduction of waviness of balls of 0.2 to 0.05 μm reduces the level of vibration by 11 db.

The relationship obtained from experiments indicates that the vibration level of bearings can be considerably reduced by making the manufacture of working surfaces of bearing components more perfect. Thus, the change of roughness of inner races from the 9th class to the 10th (according to GOST standard classification) makes it possible to cut vibrations by 3.5 or 4 db.

It is interesting to note that maxima of spectral intensity are at frequencies of 600–700, 1400–1500, and 7200 Hz; the greatest reduction of vibration occurs in the high-frequency region of the spectrum (450–9,500 Hz); no changes are observed in the low frequency region; and frequency spectra in the region of 50 to 150 Hz for almost all bearings coincide.

It is also known that the relation between the surface of races or its surface roughness and sound pressure is proportional to the sum of the product of the number of asperities by their height and by load on inner and outer rings and balls. Loading on these elements depends on ratio between the number of contacts of one point of each element with the others. For example, if the load on the outer ring equals 1, then on the inner ring it is approximately equal to 2, and on balls it equals 10. Therefore, the amount of microasperities on balls should be 10 times less than on the outer ring. Taking all these facts into account, grinding and superfinish gain a decisive importance. Ito, a Japanese researcher, has experimentally established that the fundamental frequency components of an assembled bearing are hardly changing with changing rpm, but the sound pressure level is increasing approximately by 6 db for doubling rpm.

The assembly noises arise mainly due to vibrations of the races. Consequently, the cause of assembly noise is the contact between the balls and races. Sound pressure of assembly noises is created by the radial loading applied to the groove of the ring, or by the contact loads between balls and races, and depends on the product of magnitudes of these loads by angle of the loading zone.

If we consider vibrations as a criterial parameter, it is necessary to make clear which characteristics have to be used for evaluating the level of vibrations. A straightforward record of a realization of vibrations cannot be a useful source of information due to the excessive amount of information contained. The most suitable is spectral and correlation analysis.

The first characteristic being obtained during such an analysis is the correlation function. It characterizes the degree of correlation between the values $x(t_1)$ and $x(t_2)$ of random process being apart by an interval of time $\tau = t_2 - t_1$, and is determined by two-dimensional distribution law.

Correlation function between the resistance torque of a bearing and its vibrations is decaying with increasing τ and does not depend on initial time reference. It confirms the random character of the resistance torque and vibrations of bearing rings, which can be deserted as a stationary ergodic process. In a physical sense, correlation function is characteristic of inertia of the system subjected to random excitations. In evaluating dynamic characteristics of rolling contact bearings, the correlation function is used mainly as an intermediate stage for estimation of spectral density.

The curves of correlation function in Fig. 2.13 show the correlation of the process being tested with the working conditions. In the majority of cases the correlation functions have an oscillating nature which can be caused by the presence of pronounced vibrations with discrete frequencies and with "noisy" amplitudes. The presence of noises indicated by a more rapid decay of the correlation function in the initial region as compared with the decay rate at other values of τ. The oscillating nature of the correlation function is more pronounced in the higher frequency region.

The test results show that the friction torque and 3-D vibrations of rolling contact bearings in general are stationary random functions of an ergodic type. Depending on geometry of the bearing, the characteristics of bearings also contain periodic components. Discovery of hidden periodicities, i.e., discerning of spectral structure of actual processes from the results of their measurement has a great importance for the study of vibration phenomena. Extraction of the periodic components from oscillograms allows for discovery of the basic sources of friction torques and vibrations in bearings. Spectral densities of friction torques and vibrations of bearing rings justify certain conclusions on the expected levels and frequencies for both self-contained instrument rolling contact bearings and the assemblies comprising them.

The example in Fig. 2.13 shows the test results for 840154-type bearings. These results can serve as an exemplary methodology when determining the

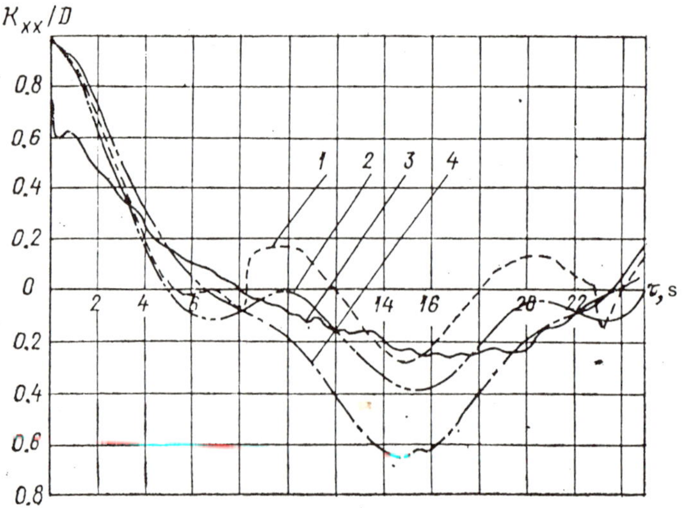

Figure 2.13 Correlation functions of VFT (*1, 2*) and vibrations of rings (*3, 4*) of a bearing of 840154-type with radial load $R = 0$, axial $A = 1$ N, $n = 2$ rpm (*1, 3*—dry bearing; *2, 4*—lubricated bearing).

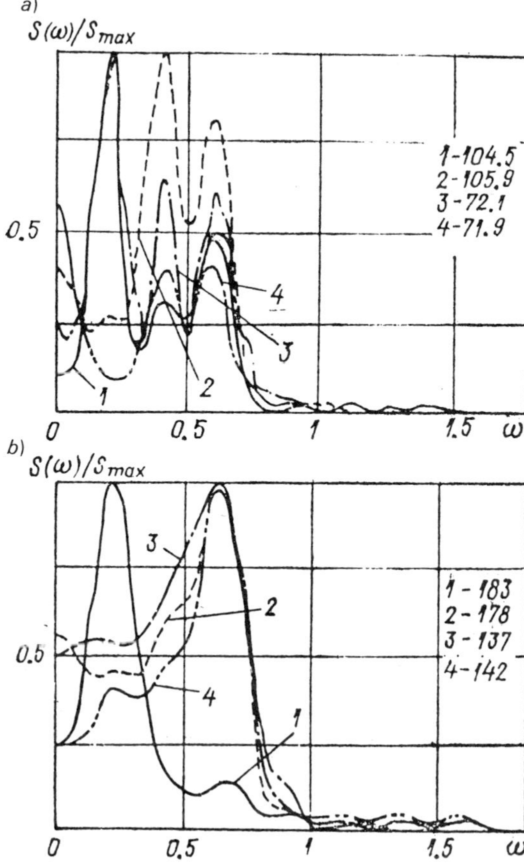

Figure 2.14 Spectral densities of axial vibrations (*a*) and resistance torques (*b*) of a bearing: *1, 2*) dry bearing rotating clockwise and counterclockwise; *3, 4*) lubricated bearing rotating clockwise and counterclockwise, respectively.

causes due to which the principal possible errors occur. The obtained spectral densities are shown in Fig. 2.14.

Analysis of the frequency composition of the process obtained in tests on a set of miniature 840154 bearings enables us to draw some conclusions on the behavior of bearings with an inner diameter 0.6–3 mm at a speed of 3×10^{-2} rps.

1) The fundamental frequency harmonic is a multiple of the number of revolutions of the outer ring of the bearing (when the inner ring is stationary);

2) The second harmonic is a multiple of the number of revolutions of the cage, and the frequency level very much depends on lubrication, installation of the bearing, and defects in the geometric shape;

3) The cage rpm depends on the axial loading and the rpm of the moving ring of the bearing (especially at low speeds);

4) The level of other harmonics is negligible.

In order to have a more complete picture of the correlation between resis-

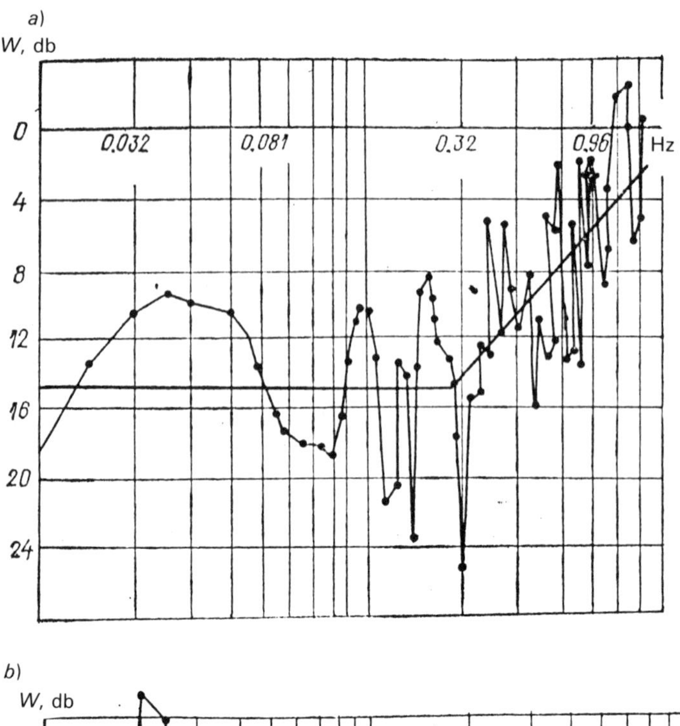

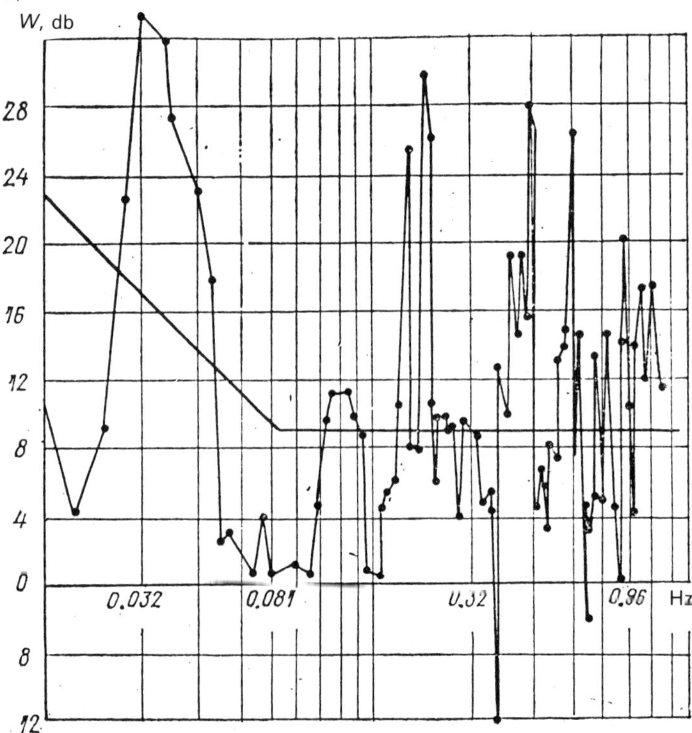

Figure 2.15 Logarithmic amplitude-frequency characteristics of: *a*) resistance torque with axial vibrations of rings, lubricated bearing; *b*) same as in *a*, dry bearing.

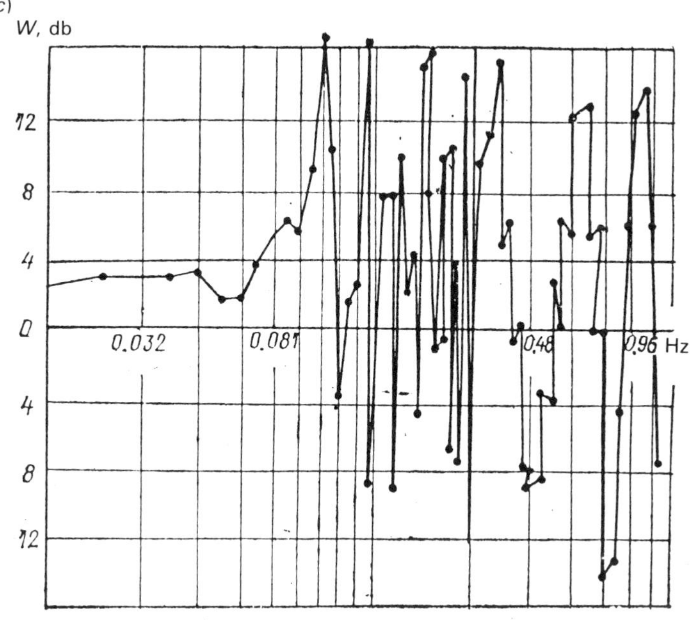

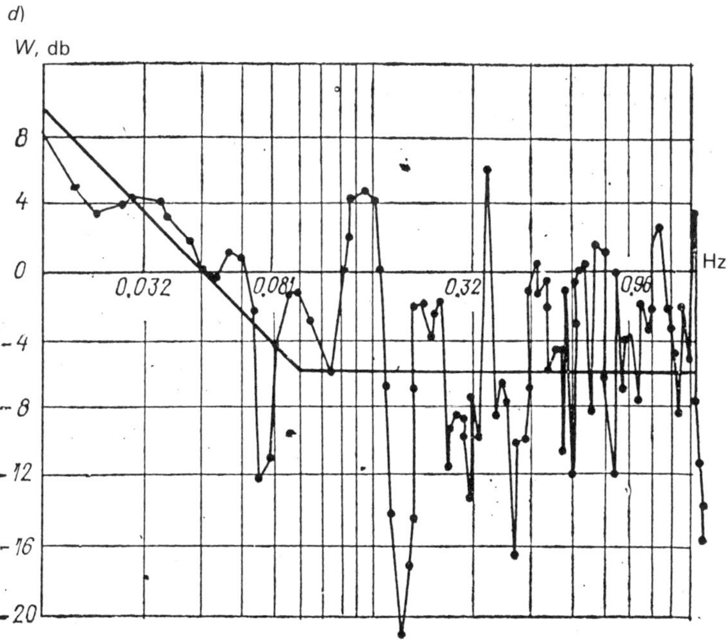

Figure 2.15 (*Cont.*) Logarithmic amplitude-frequency characteristics of: *c*) geometric defects of races, lubricated bearing; *d*) same as *c*, dry bearing.

tance torque and other parameters, for example, axial vibration of the bearing rings, geometry with respect to each other, as well as to determine variation of these correlations under different working conditions, it is convenient to use transfer functions obtained on computers.

Examples of transfer functions obtained for determining the correlations of dynamic characteristics of 840154 bearings are shown in Fig. 2.15. The plots are presented in the form of log-log amplitude-frequency characteristics. The obtained data indicate correlations between various parameters of bearings at different operational conditions. Thus, shape deviation in the form of ovality and poligonality (i.e., uneveness of race with the wavelength of more than 1/5 of the length of the circumference of the surface being machined) create large friction torques of a low frequency. Surface roughness relatively lowers the level of friction torque.

Log-log amplitude-frequency characteristics $W(j\omega)$ of a system which has at the input deviations of geometric shape and friction torque, fully coincide at the output for the cleaned (dry) bearings and they confirm the correctness of the drawn conclusions.

Correlation between geometric deviations and resistance torque obtained in the form of function $W(j\omega)$ indicates the proportional effect of geometric deviations on the resistance torque within the whole range of frequencies.

Reciprocal analysis of transfer functions of geometric shape of races of rings together with the resistance torque of cleaned and lubricated bearings confirms the hypothesis that geometric deviation with a large wavelength increases the level of friction torque more appreciably for lubricated bearings than for the dry ones.

For relatively high frequencies lubricant acts as a damper of oscillations.

Transfer functions of resistance torque and vibrations are of great interest. In the region of low frequencies deviation of torque characteristics from the characteristics of axial vibrations are not discernable. Only for the resistance torques to the higher frequency components appear which are created by surface roughness, while on the vibrations characteristics they are not observed.

Thus, analysis with the help of computers leads to the conclusion that the spectral-correlation analysis is more applicable for research purposes, while for the shop floor control of characteristics of bearings other methods are more desirable.

CHAPTER

THREE

MEANS OF QUALITY CONTROL AND DIAGNOSTICS OF ROLLING CONTACT BEARINGS

Rolling contact bearings belong to those basic components of instruments whose perfectness determines the state of the art of instruments and machines as well as automation systems. Rolling contact bearings find wide application in instruments for various fields of engineering (measurement, control, magnetic tape recording, optoelectronic systems, electrical servomotors, appliances, etc. and exert substantial effect on their accuracy, reliability, and working life.

Statistic parameters of bearings reflecting their geometric accuracy, surface roughness, stiffness characteristics of materials and kinematic couplings, do not always determine the suitability of the bearing for functioning in a specific unit or instrument. Therefore, along with verification of statistical characteristics and geometric parameters it is necessary that dynamic behaviors of the bearings be determined and studied. In the majority of cases, instead of measuring all the parameters of bearings, control of one dominating parameter may be substituted. The most informative parameter is selected for each specific case. Friction torque or vibrations of rings is often such a parameter for instrument bearings.

One of the methods for enhancing reliability of rolling contact bearings is their quality control aimed at diagnosing defects both in manufacture and operation. By diagnostics is meant determination of inner parameters of an object or their deviation from the standards by indirect indications and without disturbing the functioning of the object. Each object subjected to diagnostics has plenty of indicative signals, both informative and insignificant. Specifically, vibroacoustical processes accompanying operation of machines contain a wide range of characteristics which serve as determining factors for diagnostics. This accounts for the fact that vibroacoustical diagnostics has been gaining such rapid development in recent years.

3.1 METHODS OF QUALITY CONTROL AND DIAGNOSTICS

The increasing demands for accuracy, reliability, life, and quality of work of bearings led to more thorough testing.

The most important techniques and goals of tests of instrument bearings are presented in Table 3.1 [37].

Mode and volume of bearing tests depend on their goal and specifications. Various parameters of bearings are verified both by conventional measuring

TABLE 3.1 Types and goals of tests of instrument bearings

Type of test	Goal of test
Control of accuracy and quality of manufacture of bearing components	Determination of dimensions of balls, races, cages, deviations of balls and races from correct geometric shape, determination of surface roughness, metallographic studies, etc.
Control of geometric parameters of the assembled bearings	Determination of overall dimensions, clearances, accuracy of rotation, contact angles, axial displacement of rings, etc.
Control of dynamic parameters of bearings	Determination of bearing-generated vibration and noise, resistance torques, the remaining resource, reliability, life, etc.
Climatic, temperature, and vacuum	Verification of performance under conditions of increased humidity and temperature changes (for example, in the range of 70–140 °C), in low and deep vacuum (for example, 1.10^{-3}–1.10^{-6} G Pa)
On resistance to overloading	Verification of performance of bearings under conditions of prolonged and short-term linear vibration and impact overloading.
Fatigue resistance	Verification of fatigue endurance of bearings.

devices or setups and by specialized ones. The diagnostic techniques by which the defects are determined without disassembly are of the greatest value and deserve proper attention.

Data on temperature, the state of lubricant, power loss (friction torque), vibroacoustical characteristics, etc. may be used as parameters in judging the state of both a separate bearing and the bearing installed into a unit. The temperature conditions of bearing assemblies indicate certain defects, yet the empirical results show that the temperature of bearing assemblies is not directly proportional to their defects. This can be explained by the fact that the state of lubricant as well as its structure changes during the working process of bearing assemblies as a result of wear and accumulation of contaminants. By monitoring structural changes of lubricating oil it is possible to diagnose the degree of wear in bearings. In some cases it is helpful also in discovering the initial defects. That is why this method is frequently used in diagnosing certain defects in the bearing supports.

The test results show that defects in bearings incur a definite effect on the magnitude and character of variation of friction torque. As a result of statistical processing of experimental data it is possible to obtain correlations that allow for the nondestructive evaluation to be carried out. Such methods require:

—establish correlations as a result of statistical analyses;

—collect the necessary quantity of statistical data.

In those cases when gradual dismantling of a bearing is possible, the defects are discovered visually (though such assessment is not always sufficiently objective) or by means of roundness measuring instruments.

The methods considered here are more suited for laboratory testing because of their inefficiency. Secondly, they do not reveal all the defects in assembled bearings.

In the works [4, 34, 49, 52, etc.] on technical diagnostics, special emphasis is given to the methods of vibroacoustical diagnostics. In vibroacoustical diagnostics, vibration or noise of a bearing is looked upon as an indicating signal. Vibroacoustical diagnostics belongs to nondestructive methods of evaluation, for which bearings do not need to be taken apart. The advantage of these methods lies in the fact that by using them we practically are able to determine all possible faults in bearings whether they initiated during manufacturing, assembly, or operation.

The accuracy of assembled bearings is determined by the quality and accuracy of manufacturing of their components (rings, balls, cages).

The accuracy of a bearing assembly is dependent on the accuracy of:

—bearings, i.e., by their radial and axial runouts as well as clearances;

—machining of mounting seats for the rings (on a shaft and in a housing);

—manufacture of components designed for supporting and fixing the bearing rings;

—machining of mounting seats on a shaft and in a housing for these components;

—assembly and adjustment of radial and axial clearances, etc.

Accuracy of balls and rings of bearings as well as assembled bearings, in accordance with the recommendations of the following standards: St SEV 774-77 (Council for Mutual Economic Assistance, Comecon), STATE STANDARDS/GOST 520-71, and GOST 3722-81, are characterized by the following parameters:

For balls: by ovality, poligonality, and scatter of dimensions;

For rings (races) of bearings: by variability of a diameter of the cylindrical surface of the bearing; by variability of a width of a bearing ring; by average taper of cylindrical surface; by the runout of outer cylindrical surface (or bore) of a bearing ring with respect to the basic end face; by runout of the basic end face of inner ring with respect to bore; by radial runout of race of each bearing ring; by an axial runout of a race in a ball bearing;

For assembled bearings: by radial runout of outer (inner) ring of an assembled bearing; by axial runout of an outer (inner) ring of an assembled bearing.

Control of bearings is carried out under specified loading, at constant temperature, etc., in accordance with St SEV 774-77 and GOST 520-71.

A surface roughness of mounting surfaces of bearings is verified visually against standards or by means of measuring instruments.

Hardness, residual magnetization of bearing components decarbonization is verified in accordance with specifications. Bearings manufactured according to special technical conditions (specifications) are subjected to special control.

Schematics of instruments for controlling manufacturing errors of assembled bearings are available in reference literature. Schematics for control of bearings are given in Fig. 3.1. For measurement of non-roundness of bearing races and balls, various roundness meters are used. During the control of balls the exterior appearance, surface roughness, dimensions, tolerances, degree of accuracy, degree of poligonality, hardness, and ultimate load are checked.

3.2 MEASUREMENT OF RESISTANCE TORQUE (RT) IN ROLLING CONTACT BEARINGS

Depending on the technique of creating counteraction force, gauges for measuring RT can be divided into several different groups, and further into subgroups depending on the principles of measurement and recording of RT (Fig. 3.2).

As a rule, measurement of variable and constant components of RT can be performed by any gauge with the exception of the very first primitive ones designed only for the measurement of a constant component of RT, and also of the specialized instruments designed for very accurate determination of starting torque of a bearing or of the complete rotor system.

The simplest among these devices are mechanical devices with direct reading (Fig. 3.3). The principle of their operation is based on the balancing of

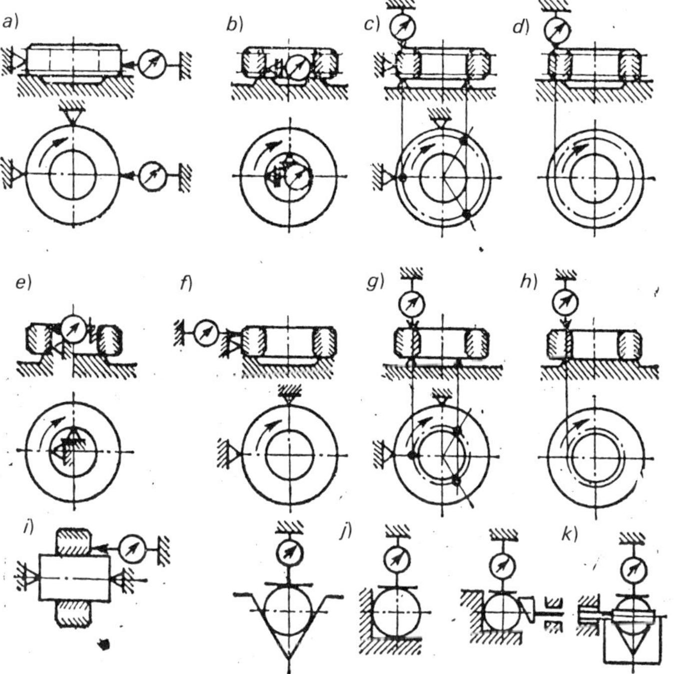

Figure 3.1 Schematics for control of manufacturing defects in assembled bearings; the parameters being controlled: *a*) diameter of outer cylindrical surface of bearing rings; *b*) bore diameter of bearing rings; *c, d*) width of outer ring of a bearing; *e*) runout of the bore of inner races with respect to the end face; *f*) runout of the outer cylindrical surface of outer races with respect to the basic end face; *g, h*) width of inner races; *i*) runout of the face of inner bearing rings with respect to the bore; *j*) poligonality and ovality of balls (the degree of accuracy 2); *k*) scatter of dimensions of balls (the degree of accuracy 02).

friction torque by the torque created by a counterweight, spring, or other mechanical system. The friction torque is computed using the expression

$$T = Pl \sin \alpha$$

where P is the mass of the pendulum; l is the distance from the center of gravity of the pendulum to the axis of rotation; α is the angle through which the pendulum swings.

Mechanical devices are suitable for shop-floor control because they do not require highly qualified personnel and have a sensitivity up to 10^{-6} Nm with an error of 10%.

In order to measure the starting torque an instrument C-21 (Fig. 3.3*h*) may be used. Using this instrument we measure the starting torque determined by the formula

$$T_{st} = (P_1 - P_2)d/2$$

where P_1 and P_2 are the masses of weights; d is the shaft diameter.

Instruments and components of a pendulum type (Fig. 3.3*a*) allow for measurement of the constant component of friction torque and also for comparative measurements of bearings in which friction torque in the bearing support is proportional to the number of oscillations of the pendulum from the start of its movement until it comes to rest or until the pendulum reaches a given amplitude φ_0:

$$T_{st} = [Pl(\cos\varphi_0 - \cos\varphi)\,57.3]/[(\varphi - \varphi_0)(3\cos\varphi_0 - 2\cos\varphi)\,2n]$$

Here P is the mass of the pendulum; l is the distance from the center of gravity of the pendulum to the axis of rotation; n is the number of periods of pendulum oscillations from the beginning of motion until it comes to rest or reaches the angular amplitude φ_0.

Determination of the friction torque of bearings by the rundown method can still be encountered in practice. The outer ring of the bearings is carefully moved by a finger. Steep slowing down or abrupt stopping indicates the presence of excessive friction torque or some defect in the rolling elements of the bearing. In this case the friction torque in the bearing can be approximately computed from the formula

$$T = 4\pi Jn/t^2$$

where J is the moment of inertia of the bearing outer ring; n is the number of revolutions; t is the time.

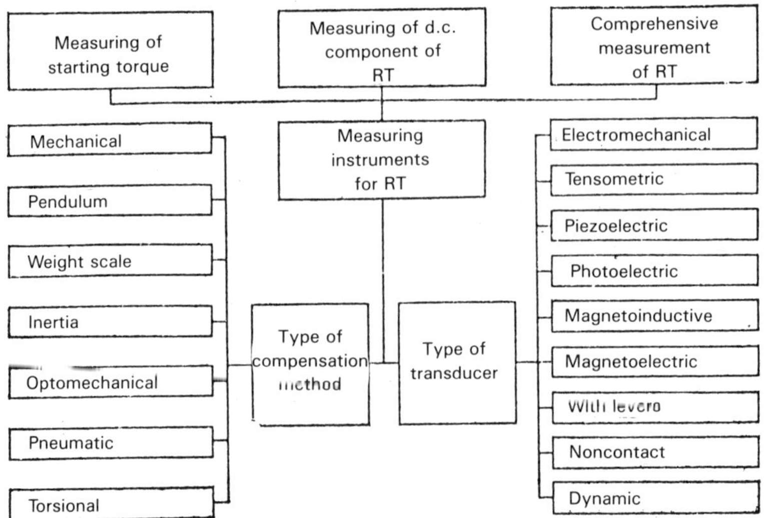

Figure 3.2 Classification of strain gauges of torques.

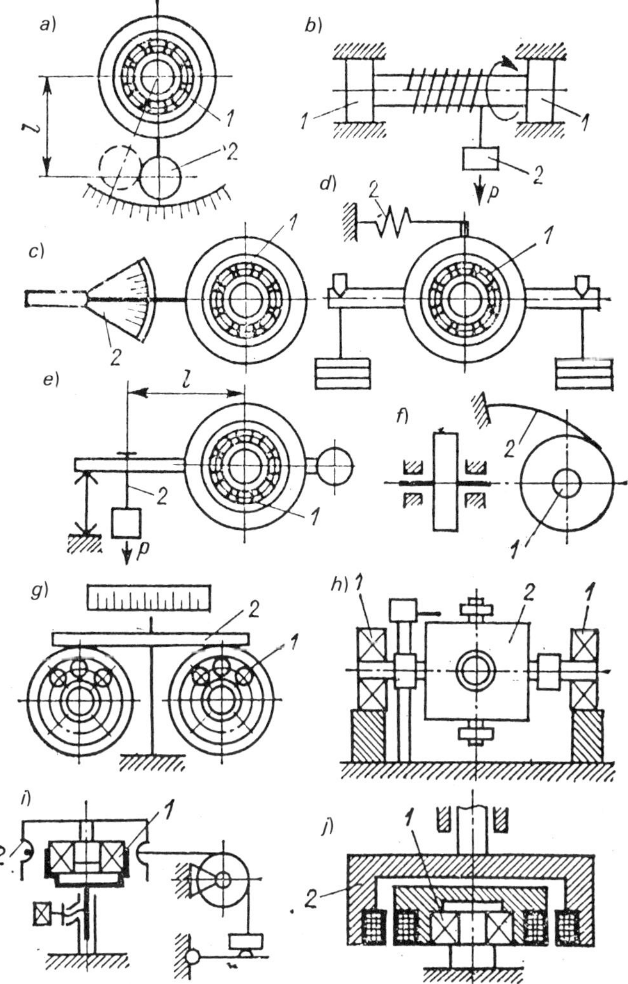

Figure 3.3 Mechanical devices for determination of friction torque: *a*) pendulum device; *b*) deadweight device (grammometer); *c*) scales; *d*) mechanical friction scale; *e*) lever-based device; *f*) spring balance unit; *g*) inertial device; *h*) instrument of C-21-type with drum; *i*) instrument for determination of friction torque by monitoring run-down; *j*) an improved device for run-down determination of friction torque; *1*) the test bearing; *2*) loading-compensating device.

In order to measure small torques, gauges with torsion bars and other elements are used successfully (Fig. 3.4). Torsion bar torquemeters enable measuring the torque down to $1 \cdot 10^{-8}$ Nm and even lower. In addition to it, microtorquemeter strain gauges, piezoelectric, variable-capacitance, and other transducers may be used. Parasitic torques in the gauges are caused by supporting units. The use of air bearing support, whose friction torque is by several orders

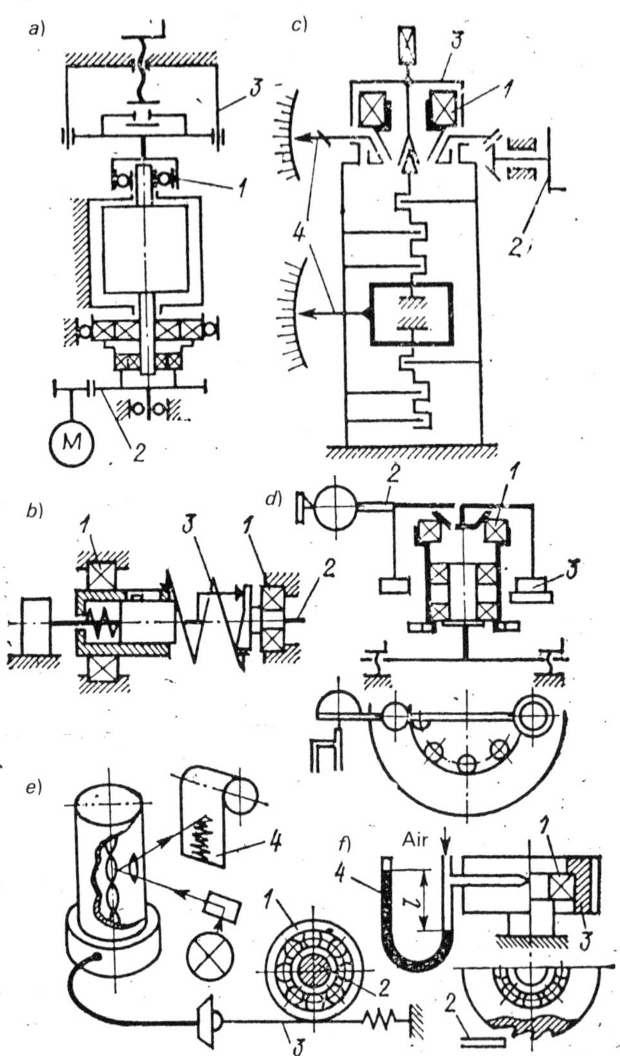

Figure 3.4 Instruments and devices for measurement of friction torque with elastic and pneumatic sensitive elements: *a*) torsional torquemeter; *b*) torquemeter with a spring sensitive element; *c*) torquemeter with tension members; *d*) MDSh-1 instrument; *e*) pneumatic diaphragm device; *f*) instrument with fluid manometer; *1*) test bearing; *2*) drive unit; *3*) loading device; *4*) indicator/recorder.

of magnitude lower than that of the bearings under test, enables to increase measurement accuracy.

The pneumatic instruments and devices have high sensitivity; one is shown in Fig. 3.4*f*. A corrugated diaphragm connected by an air conduit with a projection micromanometer serves as a sensitive measuring element in this device. Here the closed volume of air conduit limited by the corrugated diaphragm of the micromanometer and by the measuring diaphragm is changing in proportion to the load applied to the diaphragm. As a result, the air pressure changes, causing deflections of the diaphragm of the sensitive micromanometer.

Optomechanical devices and instruments form a small group, since optical systems play only the role of recording element, while the friction torque is balanced by pneumatic systems, elastic rods, etc.

Recently, the strain gauge-based methods have found wide applications in the measurement of friction torques in bearings. The basic principle of strain gauge transducers is the change in the resistance of a metal or semiconductor under the action of an applied load. The resistance strain gauges find very wide application due to their small sizes and the possibility of measuring both static and dynamic torques. The most popular are wire, foil, and semiconductor resistance strain gauges.

The relative change in the resistance of a strain gauge is determined as

$$\Delta R/R = (\Delta l/l)(1 + 2\mu)$$

Δl is the change in length; μ is Poisson's ratio.

The basic parameter of the strain gauge (gauge factor) is expressed in the following form

$$S = \Delta R/R/(\Delta l/l) \doteq \Delta R/(R\varepsilon)$$

where $\Delta R/R$ is the relative change in the resistance of the strain gauge; $\epsilon = \Delta l/l$ is the relative change in the length of the gauge.

The value of the gauge factor depends on the properties of the material the resistance strain gauge is made of and on the technology of its manufacture since

$$R = \rho\,(l/F_0)$$

Here R is the electrical resistance of the strain gauge; l is its length; F_0 is the area of its cross section; ρ is the specific resistance.

Wire resistance strain gauges are most widely used in practice. Materials having a large gauge factor and small changes in resistance with temperature are used for the manufacture of these strain gauges. For this purpose the most suitable material is constantan wire bonded onto paper on film base. The film base is prepared from Bakelite lacquer, BF-2 adhesive, or from special compositions. Films prepared from BF adhesive are suitable for use in the temperature range of $-40°$ to $+70$ °C. The Bakelite lacquer bases can work up to 200 °C.

The gauge factor of commercially available strain gauges of 2PKB and 2PKP types is 2 ± 0.2. The nominal operating current during bonding on the metal part is 30 μA. The maximum allowable strain does not exceed 0.3%. For constantan wire the temperature coefficient of resistance is $\pm 50 \cdot 10^{-6}$ deg^{-1}, the coefficient of linear expansion $\alpha = 15 \cdot 10^{-6}$ 1/°C; maximum working temperature 500 °C.

The next variation of resistance strain gauge is the foil type transducer made of thin strips of constantan foil of rectangular cross section ($b = 4$ to 12 μm). Foil-type strain gauges have a large area of contact with the object under test and hence the heat transfer is better than with wire gauges. This makes it possible to increase the supply current for the gauge and thereby increase its sensitivity.

The first semiconductor strain gauges appeared in the Soviet Union and abroad in 1955–1958. They have found very wide application.

The gauge factor of semiconductor strain gauges, unlike wire and foil-type gauges, closely depends on the level of deformation, temperature, specific resistance, and crystal lattice direction.

The dependence of the gauge factor on the temperature of the environment is determined in percentage as

$$\Delta S/S_0 = -\gamma (T - T_0)$$

Here ΔS is the absolute increment in the value of the gauge factor at a temperature T; S_0 is the value of the gauge factor at a temperature T_0; γ is the temperature coefficient of the gauge factor.

Semiconductor strain gauges are made from silicon, germanium, and such compounds as galium arsenide (GaAs), galium antimony (GaSb), and others. In germanium strain gauges a dendritic germanium strip is used with a specific resistance $\rho = 1 \pm 0.2$ Ohm · cm.

The gauge factor of semiconductor strain gauges is 50–60 times more than of wire resistance gauges. The gauge factor of semiconductor strain gauges of the KTD and KTE types with p-type conductivity is 110 ± 10 and the working temperature range lies between $-160°$ and $+300$ °C. The temperature coefficient of resistance is determined as $(5\text{–}10)10^{-4}(\Delta R/R)(1/c)$.

Force measurement with strain gauges is done by determination of deformation of the plate to which they are bonded. The resolution of such measurement is determined by an additive error created by zero drift and by noise in the measurement tract [13].

Let us consider the additive error in case of connection of strain gauges in a half-bridge circuit shown in Fig. 3.5. Here we assume that resistances of the

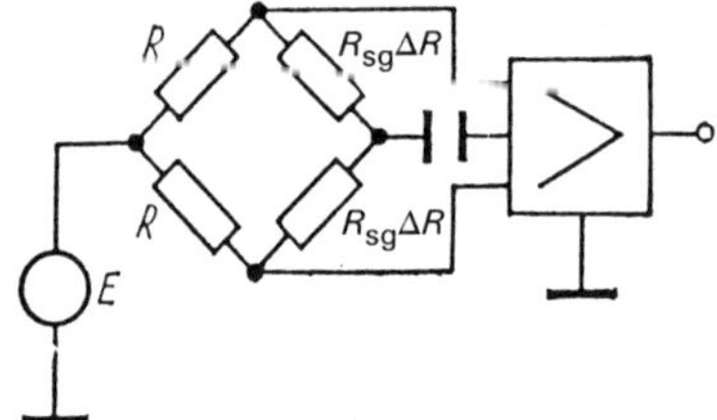

Figure 3.5 Use of strain gauge by the half bridge circuit.

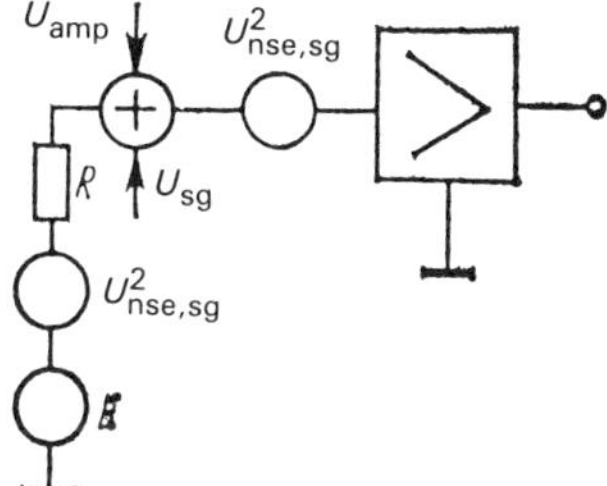

Figure 3.6 Equivalent circuit for strain gauge measuring system.

half-bridges are equal, and the half-bridges are not loaded. In order to measure the value of the additive error the circuit shown in Fig. 3.6 was used. In Fig. 3.6: $e = E\Delta R/(2R)$ is the voltage of the bridge imbalance; U_{amp}, U_{sg} is zero drift of an amplifier and strain gauge referred to the input of amplifier; $U^2_{nse,sg}$, $U^2_{nse,amp}$ are the noise voltages of a strain gauge and amplifier.

Expressing the relative change of resistance $\Delta R/R$ via force applied to the strain gauge and gauge factor, we determine the signal voltage at the input of the amplifier as

$$e = (E/2)\, K_T P \alpha \tag{3.1}$$

Here K_T is the gauge factor; P is the force; α is the coefficient depending on geometric shape and material of the plate to which the strain gauge is attached.

After assuming that the voltage of the cumulative error created by zero drift and noise is U_n, for the condition of equal values of the signal and error we find the minimum measurable force as

$$\Delta P = 2U_n/(EK_T\alpha) \quad \text{or} \quad \Delta P = U_n/K_{sg} \tag{3.2}$$

where K_{sg} is the gauge factor of sensitivity of a force transducer.

Let us consider now in a more detailed way the components of the cumulative error. Noise signals from the gauge are basically of thermal nature. Their value may be computed from Nighquist's formula, which at $T = 290\,°\text{C}$ will have the form

$$U^2_{nse\,sg} = 4 \cdot 10^{-3} \sqrt{R\,\Delta F} \tag{3.3}$$

where R is resistance of the strain gauge bridge, Ohm; ΔF is the equivalent band of amplifier noise, kHz.

The noise of the amplifier can be taken into account by the coefficient K_{nse}. It may be reduced by selection of the carrier frequency of the supply voltage of strain gauges beyond the spectral limits of the excess noise of the amplifier.

The drifts of both amplifier and strain gauge are random. The share of errors contributed by them may be reduced by an automatic correction of the drift. This is accomplished by the periodic determination of the magnitude of a drift and by its subtraction from the output signal in time intervals between the corrections. A random signal of the residual drift in this case can be considered as a periodically unstationary one with time dependent dispersion. As the work [46] indicates, dispersion of the residual drift does not exceed the value

$$\Delta_{\Sigma}(T) = 2\,\Delta\alpha\,[1 - P(T)]$$

where Δ_{Σ} is the dispersion of the drift without correction; $P(T)$ is the standard autocorrelation function of the drifting process; T is the correction period.

Reduction of the drift in this manner restricts the responsiveness of the measuring setup as well as the time allowed for interrupting the measurement in order to perform correction. The obtained value Δ_{Σ} will depend to a considerable extent on heat exchange between the strain gauge and surrounding medium.

In the long run, taking into account the possible correlation between the separate components of the error voltage U_u and assuming that the maximum magnitudes of the components do not exceed the triple values of the root-mean-square, we get

$$U_u = 3\left[\sqrt{\Delta_{\Sigma}} + \sqrt{U^2_{\text{nse}}}\right] \quad \text{where} \quad U^2_{\text{nse}} = K_{\text{nse}} U^2_{\text{nse,sg}}$$

Using the equality (3.1), from the expression (3.2) the minimum measurable load may be determined. In order to evaluate it, Table 3.2 presents some experimental data.

As it could be expected, the semiconductor strain gauges have a greater sensitivity. Their sensitivity may be as high as 10^{-5} V/N. The difference in natural frequency between strain gauges is explained by the type of bond and

TABLE 3.2 Characteristics of strain gauges

Type of strain gauge	Dimension of plate, mm	Base of strain gauge, mm	Natural resonance frequency, Hz	Supply voltage, $V \cdot 10^2$	Sensitivity, V/N
Wire strain gauge 2PKP-10-100GV	38 × 8 × 0.08	10	90	8	$8 \cdot 10^{-6}$
Wire strain gauge PKB-5-100-111	32 × 8 × 0.08	5	100	8	$8 \cdot 10^{-6}$
Wire strain gauge 2PKB-15GV	32 × 8 × 0.08	15	140	4	$4 \cdot 10^{-6}$
Foil strain gauge 2FKPA-10-100GV	32 × 8 × 0.08	10	90	8	$5 \cdot 10^{-6}$
Foil strain gauge 2FKPA-10-100	32 × 8 × 0.08	12	100	4	$5 \cdot 10^{-6}$
Semiconductor strain gauge Yu-8B-3	32 × 8 × 0.08	2	90	4	$1.8 \cdot 10^{-5}$
Semiconductor strain gauge KTE-2	32 × 8 × 0.08	10	90	4	$1.1 \cdot 10^{-5}$

insulating materials. Zero drift, a significant parameter for strain gauges, has a random character, hence for its evaluation we will use the standard autocorrelation function $p(\tau)$ and dispersion $\Delta\alpha$. Zero drift may be reduced by using a thermally insulated strain gauge or by placing it into transformer oil (Table 3.3). In dynamic measurement the oil will act as a damper. If the duration of the measurement does not exceed several seconds, then the drift during that period and its effect on the accuracy of the measurement may be neglected in computations (Fig. 3.7).

Comparing curves *2* and *3* we see that zero drift considerably depends on the supply voltage for the strain gauge. By testing it was established that from the point of view of a minimal cumulative error the voltage E equal to 1 V is the optimum supply voltage for the strain gauge bridge. If a strain gauge is placed into transformer oil, zero drift is reduced approximately 10 times (Fig. 3.8). This may be explained by the fact that the transformer oil cools down the strain gauge and reduces the influence of the environment. In steady temperature of the environment, zero drift of the strain gauge, placed into the transformer oil, has a linear character since the strain gauge slowly heats the transformer oil, thereby generating zero drift.

To calculate the minimum measurable load of a semiconductor strain gauge when the only source of its additive error is its thermal noise for $R = 10^2$ Ohm and $\Delta F = 10^2$ Hz, then from the formula (3.3) we get $U_u = 12.6 \cdot 10^{-3}$ μV. At $K_{sg} = 1$ V/N, we get $\Delta P = 1.26 \cdot 10^{-8}$ N. Some disadvantages of the semiconductor strain gauges are insufficient mechanical strength, low flexibility, and high sensitivity to light and temperature.

The components of friction torque in bearings are measured, using strain gauges in the following way (Fig. 3.9). During the rotation of the axis caused by the friction torque in the bearing, its outer ring acts through the lever on the flat spring to which strain gauges are bonded. Its deflection, proportional to the friction torque in the bearing, is recorded by the measurement instrument, which receives the signal from the strain gauges. The setup has high sensitivity but is limited in dynamic range.

There exist numerous strain gauge devices for measuring friction torque components in bearings. A flat spring made of beryllium bronze with strain gauges bonded onto it can be employed as the balancing element.

In the devices for measurement of friction torque it is necessary that loading of bearings and units by axial and radial forces should be applied in such a way that it would not cause measurement errors. One such device (Fig. 3.10) consists of unstretchable cord *2* enveloping the assembly under test and disc *3* at whose center there is an auxiliary bearing *4*. The axis of the latter is rigidly connected to the guides *5* and flexible element *6*.

The device works in the following way. The tension in the spring *6* is transmitted as radial load to the test bearing *1* through the guide *5* with fixed axis, auxiliary bearing *4*, disc *3* and branches of the cord *2*.

The auxiliary bearing *4* was tested under the oscillatory motion with very

TABLE 3.3 Test data of strain gauges

Realization	Strain gauge	Testing mode of a strain gauge	Voltage of the bridge, V	Drift dispersion of strain gauge, drift, $(\mu V)^2$
1	Semiconductor	Without thermal insulation	2.5	172
2	Same	With thermal insulation	2.5	97
3	Same	Without thermal insulation	5	900
4	Foil	Same	10	389
5	Same	Same	5	78
6	Same	With thermal insulation	10	176

small angular amplitudes about a stationary axis. It has been experimentally established that in such operational conditions the values of friction torque in rolling contact bearings are a few orders of magnitudes lower than the value of starting torque for angles of turn—about 1°. The peak value of the starting torque is reached not immediately, but gradually, after 20 or 30 minutes (Fig. 3.11).

This phenomenon takes place due to the elasticity of the contacts between the rolling elements and races and also due to the cage.

During measurement of friction torque in the device, the auxiliary bearing will rotate through an angle whose value depends on the value of the torque being measured and the size of the flat plate with strain gauge fixed to it (depending on its stiffness). During measurement of torque components of about $1.0 \cdot 10^{-4}$ Nm, by selecting the flat plate in such a way that the angle of turn of the auxiliary bearing is about 1° we get a very small value of error.

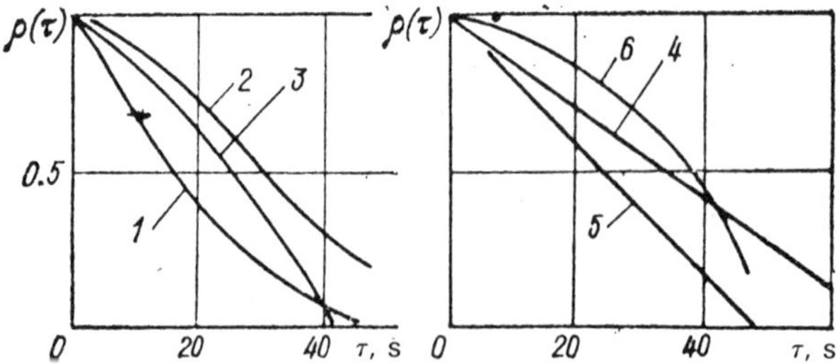

Figure 3.7 Normalized autocorrelation functions of zero drift for strain gauges: *1*) semiconductor strain gauge without thermal isolation, supply voltage for the bridge 2.5 V; *2*) semiconductor strain gauge with thermal isolation, supply voltage for the bridge 2.5 V; *3*) semiconductor strain gauge without thermal isolation, supply voltage for the bridge 5 V; *4*) foil strain gauge without thermal isolation, supply voltage for the bridge 10 V; *5*) foil strain gauge without thermal isolation, supply voltage for the bridge 5 V; *6*) foil strain gauge (same as 5), supply voltage 10 V.

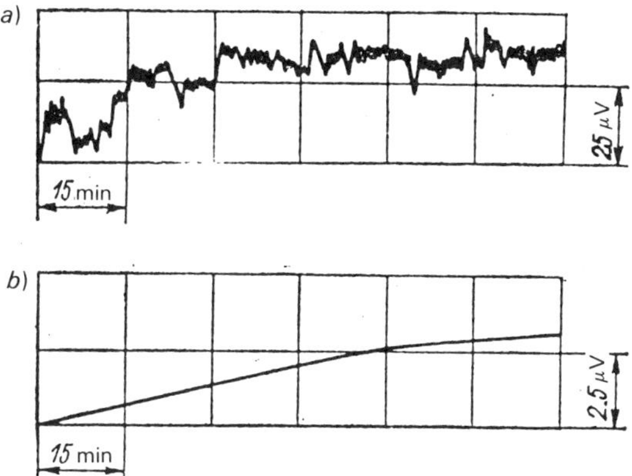

Figure 3.8 Zero drift of semiconductor strain gauge with supply voltage for the bridge 1 V: *a*) without thermal isolation; *b*) placed into transformer oil.

Another type of instrument for measurement of friction torque with elastic rods is based on using piezoelectric effect. The piezo effect is linear, i.e., mechanical deformations of the crystal give rise to directly proportional electrical polarization. The measuring instruments using piezotransducers allow measurement only of the variable component of the friction torque in bearings.

The piezotransducers have inadequate directional properties. Hence for the direct transfer of torque to the piezotransducer from the moving part of the instrument, with the inescapable runout and vibrations of the latter, considerable tangential forces act on the transducers and result in errors which occasionally even exceed the torque being measured. To avoid this effect, the system in Fig. 3.9*b* is used. This system is analogous to the system used in the strain gauge devices for a rigid transfer of force, which in essence consists of the use of a light steel needle and spherical thrust pads made of hard stones fixed to both the transducer and the moving part of the instrument.

A special type of electromechanical means for measuring friction torque components in bearings is the photoelectric method in which torque characteristics are registered by photoelectric transducers. The balancing torque can be produced by a magnetoelectric device, electric spring, induction mechanism, etc.

For example, in the case of an inductive mechanism, the friction torque is compensated by the torque $T_B \equiv c(f/\rho)I_1^2 \sin \psi$, where c is the coefficient of proportionality; f is the frequency of alternating current; ρ is the specific electrical resistance of the material of the disc; I_1 is the current in the primary winding of the transformer; ψ is the phase shift between magnetic fields. If the condition $I_1 = KI_2$, where I_2 is the current in the secondary winding, is fulfilled, then the torque produced by the inductive mechanism can be determined from the reading of the ampermeter that measures the current in one of the windings. The value of the current needed to produce compensating torque is automatically maintained by the undercompensation controller consisting of lighting arrangement, photoresistor, and amplifier.

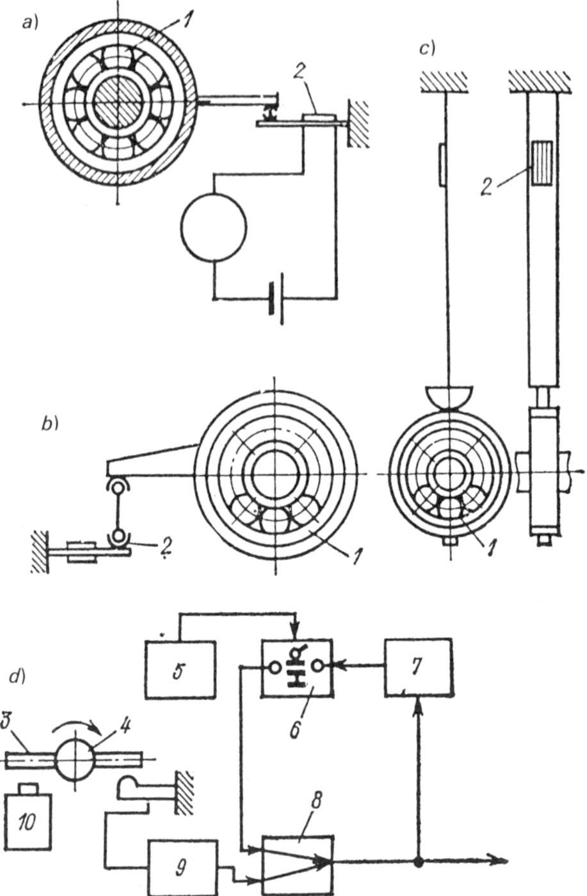

Figure 3.9 Strain gauge devices for measurement of friction torque: *a*) friction torque measurement device; *b*) improved strain gauge device; *c*) device with vertical flat spring; *d*) strain dynamometer of periodical action; *1*) test bearing; *2*) sensitive element; *3*) electromagnetic mechanism; *4*) strain dynamometer; *5*) control block; *6*) capacitor-based memory cell; *7*) divider of voltage; *8*) differential amplifier; *9*) strain amplifier; *10*) electromagnet.

Dynamic measuring instruments for torques are used for determination of both the friction torque of bearings and the driving torque, since the measurement as a rule is carried out without compensation of the torque under test by counteracting torque (USSR certificates of invention 340918, 509811, and 549696).

There are numerous other arrangements (Figs. 3.12 and 3.13) which are constructed by the same classical approach; the torque is balanced with counteracting force which can be produced by different methods (mechanical, electrical, magnetic, etc.). The registration of oscillations of the torque, i.e., its vari-

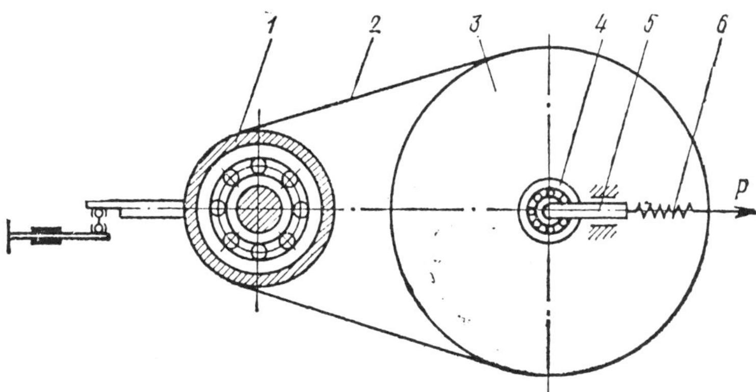

Figure 3.10 Torquemeter with the use of strain gauges: *1*) test bearing; *2*) cord; *3*) disc; *4*) auxiliary bearing; *5*) guides; *6*) flat spring.

able component, is also carried out by different types of transducers. The constant (d.c.) component of torque is generally determined from the balancing torque.

3.3 MEASUREMENT OF VIBRATIONS OF BEARINGS

The measurement of vibrations in bearings is a complex problem and includes measurement of many parameters: frequencies of natural vibrations, frequencies of fundamental vibrations, amplitudes and peak values of vibrations, displacements, velocity and acceleration, etc. The object of the measurement is the whole unit in which the major sources of vibrations are bearings, the housing of bearings, or only some elements of bearings, for example, inner and outer races and balls (in the case of ball bearings). The direction of the vibration measure-

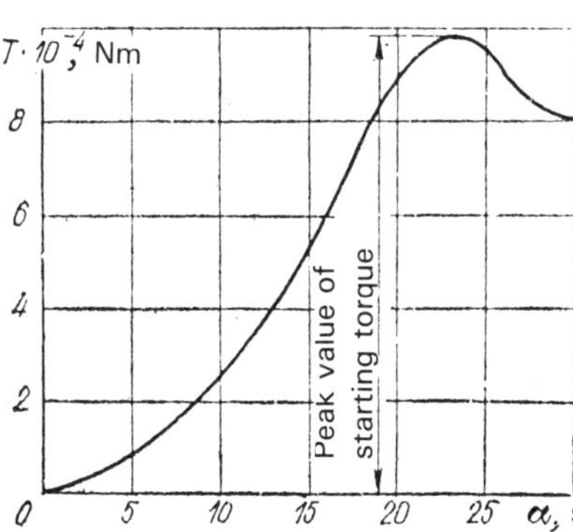

Figure 3.11 The friction torque of C1000086-type bearing.

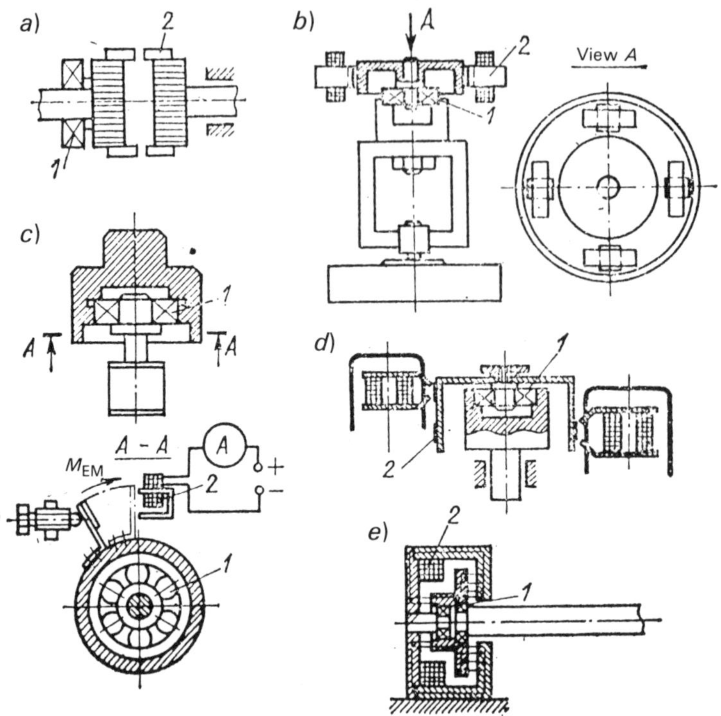

Figure 3.12 Magnetic action devices for measuring friction torque in bearings: *a*) block diagram of a magnetic unit; *b*) devide for quality control of bearings according to d.c. component of friction torque; *c*) magnetoelectric device for quality control of ball bearings; *d*) magnetoinductive setup for measuring friction torque in bearings; *e*) magnetic torquemeter; *1*) test bearing; *2*) loading-compensating unit.

ment is also important, since at different directions the levels and the frequency spectrum will also be different.

There are two methods of vibrations measurement. The kinematic method considers the motion of a rigid body, i.e., takes a reading of coordinates of separate points in the coordinate frame x, y, z. This method determines the coordinates of vibration point and angles between the axes of the stationary system and the system attached to the body. This allows for determination of the momentary position of the point for both small or large displacements in space. The basis of the dynamic method is the formation of an artificial stationary reference system. An inertial element with a considerable mass is attached on the soft spring to the point of interest of the vibrating element (elastic suspension). After damping of free oscillations, the inertia element may be considered to be at rest and the displacement of the vibrating body with respect to the inertia element can be measured. It is impossible to attain complete immobility of the inertia element. Therefore, the relative oscillations between the inertia element and the solid body represent the absolute oscillations of the body only approximately, with a certain error.

There are instruments for measuring mechanical oscillations: 1) in relation to stationary coordinates (SC); 2) inertial action (IA), measuring the vibration with the respect of the inertia element.

For the study of oscillatory processes the following techniques of photography can be used: single momentary photography suited for impact processes; multiple momentary succession; with simultaneous shifting of a photographic film.

The mirror methods are based on the measurement of deflection due to vibration of a light ray reflected from the object being tested or from the mirrors or prisms installed on the object. The rays reflected get to the scale with the element attached to it or strike the rotating drum containing the photographic paper.

The interferometric methods are based on observation of an interferometric image which is obtained as a result of interaction between two rays of monochromatic light, direct and reflected from the mirror fixed on the vibrating object.

For the measurement of vibrations by nonelectric methods vibrographs and vibrometers are often used. At present these instruments are being replaced with less complicated vibration transducers, which may be divided into generating and parametric transducers. The generating transducers are transducers with direct transformation of mechanical oscillations into electrical signal. The parametric transducers are those in which mechanical oscillations modulate

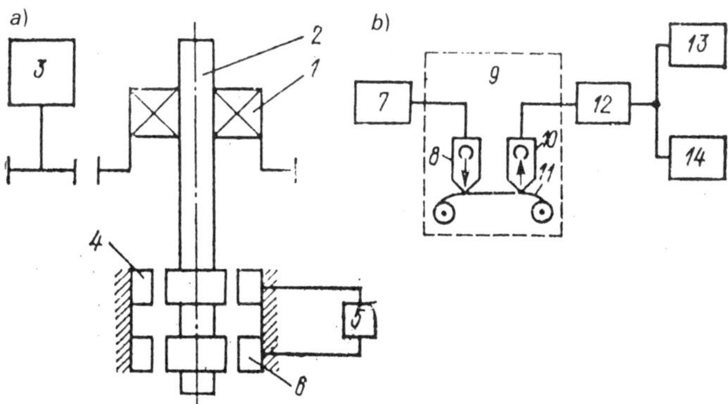

Figure 3.13 Instruments for measuring bearing characteristics: *a*) basic diagram of instrument measuring dynamic characteristics of bearings; *b*) contactless measuring instrument of the torque: *1*) test bearing; *2*) axis of transducer rotor; *3*) electrical motor; *4*) signal transducer; *5*) amplifier; *6*) torque gauge; *7*) converter; *8*) recording head; *9*) apparatus of magnetic recording; *10*) reproduction head; *11*) magnetic carrier; *12*) demodulator; *13*) computer; *14*) video monitor.

electrical sinusoidal oscillations or direct current is produced in the apparatus independent of vibrations.

The most convenient are transducers in which mechanical oscillations are transformed into electrical signals. Thus, electrodynamic, electric, inductive, piezoelectric systems and systems with modulated resistance and capacitance have found the widest application.

Instruments for the monitoring of vibrations in bearings are used for the registration of forced elastic oscillations of various components of bearings during their motion. The spectrum of bearing vibrations ranges from several Hertz to hundreds of kilohertz. The measurement is usually performed in relative units (decibels) corresponding to a change of vibration level by 1.012 times. The level of vibrations (db) $B = 20 \lg (a/a_0)$, where a is the acting acceleration; a_0 is the reference level of the acceleration. The acceleration equal to $3 \cdot 10^{-6}$ m/s^2 is taken as a reference level of acceleration during the vibration measurements.

Vibration measurements can be divided into contact and contactless (Fig. 3.14). These two groups of measurements are subdivided into other independent groups according to the types of transducers employed. We will make a short survey of some methods used in measuring vibrations.

During the measurement of vibrations with the help of the resistance transducers, the displacements and frequencies are recorded by contact methods using rheostatic transducers and transducers using contact resistance. Electrolytic, electronic, and ionic techniques of measurement should also be included into this group.

Inductive measuring techniques are based on the change in inductance under the action of input signal, in this case, displacement.

Capacitance methods are based on the change in capacitance due to variation of input parameter. As is well known, the capacitance of a condenser is

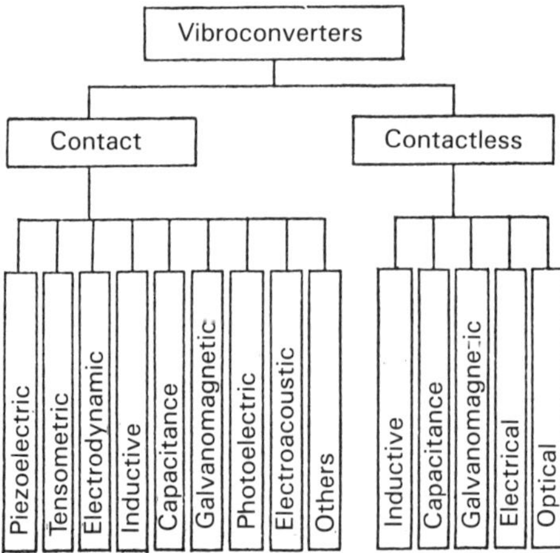

Figure 3.14 Vibrotransducers for measuring vibrations.

$C = \epsilon S/\delta$, where δ is the clearance between electrodes; S is the area of the plates; ϵ is the dielectric constant.

Radiation techniques are based on changes in the radiational energy as a result of displacement. Visible light is most frequently used. In cases where the displaced element is in a closed container, penetrating radiation is used—X-rays, γ and β rays. In certain cases it is convenient to use α radiation instead of light rays. Using various forms of radiational energy and different techniques for measuring the intensity of flow it is possible to obtain the required characteristics and properties of transducers.

Thermal methods are based on changes in electrical properties due to the effect of the temperature of the element. The dependence of heat transfer on different factors is determined by similarity criteria: $Nu = \alpha e/l$—Nusselt number; $Pr = c_p \mu/l$—Prandtl number; $Gr = \beta g l^3 \Delta t \rho^2/\mu^2$—Grasshoff number; $Re = vlg/\mu$—Reynolds number, where α is the heat transfer coefficient; l is a specific length; e is the coefficient of heat conductivity; c_p is the specific heat; μ is the coefficient of dynamic viscosity; β is the coefficient of volume expansion; v is the velocity; g is acceleration due to gravity.

Pneumatic and hydraulic methods are based on changes in pressure in the system due to vibrations. These methods are simple, have a considerable power output, and are explosion proof.

In the computation of pneumatic systems for the subcritical region of flow we use the formula

$$G = \mu F \sqrt{2g(RT)} \sqrt{p_1(p_1 - p_2)}$$

where G is the weight of air discharged; μ is the discharge coefficient; F is the area of orifice; R is the gas constant; T is the absolute temperature; g is the acceleration due to gravity; p_1 and p_2 are the pressures before and after compression.

For hydraulic systems we use the formula

$$G = \mu F \sqrt{2g/\rho} \sqrt{p_1 - p_2}$$

where ρ is the viscosity of a liquid.

Compensation methods are based on the comparison of input value with the analogous compensating quantity. Compensation systems are closed systems and are divided into static and astatic.

Vibrations are especially undesirable in precision devices, navigation instruments, and the like. Therefore, we will consider in a more detailed way the methods and means for measurement of vibrations in rotors of gyromotors rotating in ball bearings. Nowadays basically angular ball bearings are used in gyromotors. The angular ball bearing excites axial and radial vibrations of the rotor due to which the deviation and drift of the gyroscope take place. The complexity of oscillation phenomena taking place in the rolling contact bearing and their dependence on a great number of factors (rotational speed of load, perfectness of surfaces, lubricant, temperature, nonlinearity of radial clearance) present certain difficulties for accurate description of physical phenomena and leads to an experimental solution of the problem. There is still no methodology developed which would allow to state which parameters of rotors should be controlled and within what limits. Usually, the rotor vibration is characterized by the root-mean-square value of the amplitude of vibration acceleration, which is only an approximate evaluation of dynamics of a rotor.

The following requirements must be met by the experimental setup designed for correct evaluation of the effect of vibrations of ball bearings on stability of a rotor: no vibration produced by the drive of the rotating ring of the test bearing; the measurement of displacements and vibrations by the contactless technique without formation of disturbance from the slip rings; high sensitivity; thermostability of the transducers used; the rotor of exceptional stiffness.

In practice, for the measurement of vibrations piezoelectric transducers are most frequently used. In recent years some researchers used capacitance displacement transducers for the measurement of oscillations of a rotor in ball bearings. There are numerous apparata available for the measurement and analysis of vibrations of rolling contact bearings (Fig. 3.15).

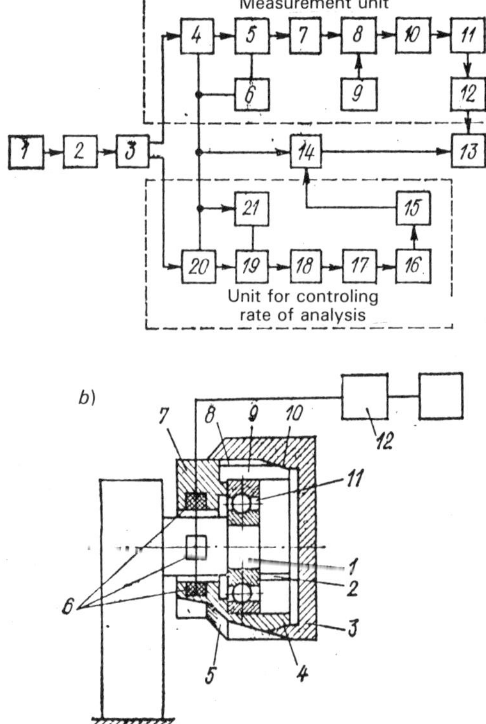

Figure 3.15 Vibration measurement of rolling contact bearings: *a*) block diagram of analyzer of accelerated action: *1*) source of information; *2*) memory; *3*) buffer amplifier; *4, 20*) narrow band device; *5, 8, 19*) mixer; *6, 9, 21*) heterodyne; *7*) intermediate frequency analyzer (IFA) 1; *10*) IFA 2; *11*) quadratic detector; *12*) low pass filter (LPF); *13*) recorder; *14*) frequency tuning; *15*) block for controlling rate of frequency tuning; *16*) communicator; *17*) amplitude selector; *18*) detector for LPF. *b*) device for measurement of vibrations of rolling contact bearings. *1*) driving shaft; *2*) sleeve; *3*) cap; *4*) cap housing; *5*) dial; *6*) vibrotransducers; *7*) sensitive element; *8*) bushing; *9*) clamping collet; *10*) collet housing; *11*) test bearing; *12*) recorder.

In the work [5] the methodology for the study of oscillations of the rotor bearing on the gas static bearing under the rotational speed 1000 s^{-1} in a horizontal and vertical position is given. Displacements of the rotor were measured by four inductive transducers. Their readings were recorded by an oscillograph.

3.4 DETERMINATION OF OIL FILM PARAMETERS

An oil film forming between the rolling bodies and races of the bearings is closely correlated with the geometric shape of bearing components and their operational conditions. The determination of the parameters of oil film should be based on these conditions.

Bearings essentially operate in dynamic conditions, hence we are mostly concerned with the dynamic parameters of an oil film, i.e., with its variable component, which as it has been established is related to the friction torque in bearings.

Analytical determination of the parameters of an oil film of bearings can be based on the geometric shape of ball races of bearings. The geometric defects in a ball race can be expressed in terms of a Fourier series.

In the case of the ball race of the inner ring the error function is

$$F_1(\varphi) = R_{0i} + \sum_{k=1}^{m} A_k \cos(k\varphi + \varphi_k)$$

and for the ball race of the outer ring

$$F_2(\psi) = R_{0o} + \sum_{k=1}^{m} B_k \cos(k\psi + \psi_k)$$

Here m and k are numbers of harmonics; A_k and B_k are the amplitudes of the harmonic components; φ_k and ψ_k are the corresponding phase angles; R_{0i} and R_{0o} are the radii of inner and outer rings of ideal geometric shape.

From the hydrodynamic theory of lubrication it follows that the thickness of the film is

$$h_{i,o} = 4\pi^2 v_{1,2}^2 \mu^2/[\beta(2\beta + 8\alpha_{i,o})^2 P_i^2]$$

where $h_{i,o}$ is thickness of the film between the ball and the inner, outer rings; $v_{1,2}$ is the speed of a longitudinal movement of the rolling point with the inner and outer rings; μ is the viscosity of the oil under atmospheric pressure; $P_i = P_1(\cos\alpha_i)^{2/3}$ is the load acting on the i-th ball; P_1 is the load acting on the most loaded ball; α_1 is the central angle between the vertical axis and radius passing through the center of the i-th ball; β and $\alpha_{i,o}$ are the geometric coefficients; $\alpha_o = 1/2[1/r - 1/(R_2(\psi))]$; $\alpha_i = 1/2[(1/r - 1)/(R_1(\varphi))]$; $\beta = 1/2(1/r - 1/R_{rc})$; r is the radius of the ball; R_{rc} is the radius of the race in the plane perpendicular to the direction of rolling.

By the methods of calculus we find the radius of curvature at the given point:

$$R_1(\varphi) = \left\{ R_{0i} + \sum_{k=1}^{m} A_k \cos(k\varphi + \varphi_k) + \left[\sum_{k=1}^{m} kA_k \sin(k\varphi + \varphi_k) \right]^2 \right\}^{3/2} \Bigg/ \left\{ R_{0i} + \sum_{k=1}^{m} A_k \cos(k\varphi + \varphi_k) + 2\left[\sum_{k=1}^{m} kA_k \sin(k\varphi + \varphi_k) \right]^2 + \left[R_{0i} + \sum_{k=1}^{m} A_k \cos(k\varphi + \varphi_k) \right] \times \left[\sum_{k=1}^{m} k^2 A_k \cos(k\varphi + \varphi_k) \right] \right\}$$

$$R_2(\psi) = \left\{ R_{0o} + \sum_{k=1}^{s} \beta_k \cos(k\psi + \psi_k) + \left[\sum_{k=1}^{s} k\beta_k \sin(k\psi + \psi_k) \right]^2 \right\}^{3/2} \Bigg/ \left\{ R_{0o} + \sum_{k=1}^{s} \beta_k \cos(k\psi + \psi_k) + 2\left[\sum_{k=1}^{s} k\beta_k \sin(k\psi + \psi_k) \right]^2 + \left[R_{0o} + \sum_{k=1}^{s} \beta_k \cos(k\psi + \psi_k) \right] \times \left[\sum_{k=1}^{s} k^2 \beta_k \cos(k\psi + \psi_k) \right] \right\}$$

Substituting into the formula for $h_{i,o}$ the values $R_1(\varphi)$ and $R_2(\psi)$ we will find the oscillations of the thickness of hydrodynamic film taking into account the defects of the races.

The power consumed for the rolling of one ball

$$U_{i,o} = (12\pi v^2 \mu \sqrt{\alpha_{i,o}})/(2\beta + 3\alpha_{i,o}) \sqrt{\beta} \arcsin h(m\sqrt{\beta}/h_0)$$

where m is the width of the oil layer; h_0 is the specific thickness of the oil layer.

The moments of forces of hydrodynamic friction in rolling of one ball at the inner and outer rings are

$$T_i = U_i/v_1; \quad T_o = U_o/v_2$$

The overall hydrodynamic friction torque in the bearing

$$T = \sum_{i=1}^{z} (T_{ii} + T_{oi})$$

where z is the number of balls.

From here it follows that the variable resistance torque depends not only on the height constituted by macrogeometric defects (nonroundness). It also depends on the radii of curvature formed by these defects. Here is a possibility of changing the frequency of the variable component of friction torque which is related to the losses associated with shear deformation in the lubricant layers.

The greater the variety of the radii, the higher is the frequency of the variable part, and vice versa.

In recent years most researchers prefer resorting to empirical methods while determining the parameters of oil film. A lot of methods are known for the measurement of thickness of oil film in a bearing. The basic ones are: X-ray; electrical resistance and voltage drop; capacitance; mechanical. There are also other techniques such as optical, pneumatic, and photoelectric. They are suited, however, mostly for the study of oil film in sliding friction bearings that have relatively large sizes and average rotational speeds.

The following conditions are to be met in design and manufacture of instruments for determination of thickness of oil films forming in high-speed ball bearings between the rolling elements and the races:

—the dial of an instrument must be linear;

—interference from the outward magnetic fields should be eliminated;

—independence of the chosen measurement technique from the effect of the temperature of the environment;

—the selected method and design of the apparatus should not affect the lubricating, chemical, and structural properties of the lubricating film.

In spite of a large amount of work on the measurement of the film thickness carried out both in our country and abroad, a really suitable technique for determination of this thickness is yet to be developed, especially when the bearings work under special conditions.

In experimental studies of bearings, the method of electrical resistance has found wide application.

As early as 1941 L. V. Yelin pointed out that electric resistance of thin lubricating films was directly proportional to the film thickness. During the measurement of electrical resistance of thin lubricating films, however, large errors arise. This is due to the fact that the measurement is carried out by direct current and thus an electrical dissociation of the lubricating oil is possible. Some additional conditions should be met in design and manufacturing of the apparatus for the measurement of electric resistance in ball bearings:

—the measurement of electric resistance has to be carried out under the alternating current;

—the drop of voltage of the resistance being measured should not exceed approximately 5 mV.

At present nearly all techniques of measuring electric resistance are based on higher frequency power supply. The measurement of the film thickness by the method of its overall resistance has been suggested by Yu. N. Drosdov and V. R. Reshschikov. It has been established that within the limits of film thickness up to 5 μm the dependence of current on the voltage applied to the lubricating film obeys Ohm's law. Changes in temperature and voltage have a certain effect on changes in electrical properties of lubricating oil. In order to eliminate this phenomenon, some researchers have put a few percents of sodium-petrolic sulphanate as additives. To the same end, other researchers have used water.

The use of additives, however, makes the measuring of films with the thickness of 1 μm more difficult because of high conductivity and formation of a two-phased region at the contact.

The technique of active resistance, which is very simple in itself, is prone to interferences. The very measurements are not complicated, comparatively, in a bearing working in natural conditions under rotational speed 0.1 to 0.5 $\times$ 10^{-2} s^{-1}. At high angular speeds, interference produced by slip rings is considerable.

The practice of laboratory testing shows that the technique of determining layer thickness by voltage drop in that layer and under the known current offers the highest accuracy and stability. This technique is also suitable for d.c. supply voltage. L. V. Yelin was the first to point out that there exists the direct relationship between the voltage drop and thickness of the lubricating layer.

In order to measure the thickness of a lubricating film some researchers have used the voltage drop under 1 A current passing through the film. The authors of the work [23] have suggested an improved method for measuring the thickness of a lubricating layer. It is based on measuring the voltage drop under the known current. A slightly different scheme is presented in [24] for the measurement of a lubricating layer forming between the two surfaces in contact. He suggested the calibration technique which is based on preliminary degreasing of the surfaces. After that, the two surfaces are brought together by turning a micrometric screw, with the reading of the indicator, millivoltammeter, and oscillograph being simultaneously recorded. Then one roller is withdrawn to a certain distance and the lubricating oil gets into the clearance.

At the central interdepartmental vibrotechnical research laboratory, Kaunas Polytechnic Institute, an investigation of the effect of hydrodynamic action of the lubricating oil film (HLF) due to its thickness on parameters of the lubricating layer (PLP) was carried out. The experimental study was performed by V. I. Dovidenas and V. N. Augutis using the method of measurement of electrical resistance of HLF in the rolling contact bearing. The bearing was considered as an assemblage consisting of a great number of cells in which the resistance and capacitor are connected in parallel. The source of voltage with a series high resistance are causing a constant current through the bearing. Measuring the drop of voltage it is possible to determine its resistance (conductance) r. Since r varies within small limits, the voltage at the bearing is amplified and recorded. If the measurement is performed by means of alternating current, then the detector must be used.

A change in electrical resistance also may be determined by the capacitance method. The bearing is looked upon as a capacitor with its active resistance being neglected. This is correct because the measurement is carried out on the frequency f = 100 kHz, thus the capacitance would shunt the active resistance. The frequency of the measuring circuit is selected so that the change in voltage on the circuit as well as the reading of the oscillograph were proportional to the change of capacitance of the bearing.

Of special interest is the attempt at measuring the thickness of the lubricating film of the rotating ball bearing and verification of the results of hydrodynamic theory of lubricant by mechanical means. The hydrodynamic theory of lubrication developed by P. L. Kapitsa states that in a working rolling contact bearing a hydrodynamic pressure of the film develops between the rolling bodies and races which tears the rolling bodies away from the races. These considerations have served as a base for development of the specialized device mentioned above.

An X-ray technique for measurement of thickness of a lubricating layer between the discs in contact is used, essentially, in specialized friction testing machines. In these cases, radiation techniques have a number of advantages over other techniques. The gist of this technique lies in the principle according to which a ray of a square cross section of parallel monochromatic X-rays of high power passes between the two surfaces of rolling rollers. The intensity of radiation which passes between the discs parallel to the planar area of the contact surfaces depends on the thickness of the lubricant layer which separates these layers. The wave length of X-rays is selected such that the rays pass without hindrance through the lubricating substance without penetrating into steel. Application of the X-ray technique for measurement of the lubricating layer directly during the rotation of the rolling contact bearing presents certain difficulties. This is the major disadvantage of the technique which restricts its practical application. Therefore it was suggested [54] to carry out the measurement on a precision roller contact machine where the rolling speed, voltages, and temperatures were kept similar to the actual working conditions of those in the rolling contact bearing. This is of course an indirect method of measuring the thickness of a lubricating layer, which explains its infrequent use.

The work [54] presents experimental data on correlation of the lubricating layer thickness with temperature, rotational speed, and the contact voltage. The results are compared with theoretical predictions.

3.5 MEASUREMENT OF TEMPERATURE OF BEARING UNITS

In bearings, especially in high-speed ones, the power loss due to friction in the majority of cases causes considerable increase in temperature at the contacts. As a rule, this results in appreciable reduction of working life and reliability of a bearing. According to the exponential law, with the increase in temperature, longevity of plastic lubricants decreases, viscosity of materials, thickness and strength of boundary and hydrodynamic films of the lubricant diminish, and the contact endurance and wear resistance of the working surfaces are reduced. With the increase in temperature, the coefficient of friction either falls or rises depending on the friction condition. This in turn changes the power loss due to friction and causes sliding with intensive wear and tear or facilitates formation of cracks, pitting, and reduces the working life of the component.

Thermal deformations of bearing units and zones of the contact cause redistribution of loads on bearings and change their stiffness and accuracy of rotation. This creates a great need to determine and control the temperature pattern, especially the temperature of the working zone of a bearing. However, high variable gradients of temperature, small-sized fast moving sources, and limited access to them for testing, together with complicated boundary conditions of heat exchange process during the theoretical study, create great difficulties for solution of thermal problems and make their practical application more complex.

At present, depending on specific conditions of the study, many methods for temperature measurement of the solid body surfaces are used. Numerous designs for measurement of differential temperatures have been developed. Additional circuits and devices are used to minimize disturbances from the outside noise sources. The most popular method for the measurement of temperature with the help of thermocouples is based on the existence of a certain relationship between the thermoelectromotive force and temperature in the contact between heterogeneous conductors. A differential thermocouple is used for the measurement of the difference of temperatures. A connection in a series of several thermocouples is used in order to increase thermoelectromotive force.

During the measurement of thermoelectromotive force by a millivoltmeter, the force of the current passing through the circuit under measurement depends not only on the thermoelectromotive force being measured, but also on the resistance of a thermocouple (and connecting wires), which vary in accordance with the wear of the thermocouple and temperature of the environment. These errors are eliminated by the compensation method when the thermoelectromotive force of the thermocouple is compared with the known difference of potentials from calibrated resistance through which a specific constant current is passing.

In self-balancing potentiometers where the balancing voltage is taken from the diagonal of a bridge, the balancing voltage changes linearly depending on the movement of the measuring potentiometer.

For the majority of automatic monitoring and recording instruments, the statistical errors are more essential. Accurate measurement of small differences in temperatures with the help of differential thermocouples may be ensured only under their individual calibration and introducing corrections into a reading of instruments. Nonlinearity of temperature, characteristic of a differential thermocouple, may be balanced by a parallel connection to the measuring device of a correcting temperature-sensitive resistor with negative temperature coefficient. This method, however, reduces sensitivity of a thermocouple and in addition to it, semiconductor temperature-sensitive resistors cannot be used under high absolute temperatures. Nonlinearity of the characteristic of a thermocouple may be balanced by means of the third balancing thermocouple.

In addition to the method of thermocouples, in order to measure temperature it is possible to use resistance thermometers, whose operation is based on the properties of metals to change their electric resistance with a change of temperature. The difference of temperatures is measured by resistance thermometers which are included into the differential measuring circuit. For the automatic measurement of the temperature difference, electronic balancing bridges are often used. Standard resistance thermometers are large, which limits their use.

In recent years, semiconductor temperature-sensitive resistors have found wide application for temperature measurement. However, due to specific characteristics and drawbacks associated with the technology of manufacturing temperature-sensitive resistors, they are not recommended when a high accuracy measurement is required.

For the measurement of temperature of solid bodies by the contactless technique, the optical (infra-red) radiation methods are used. The radiant thermometers, however, are far from perfect: their shortcomings are related to selectivity of measurement and absorption; the optical systems have unavoidable shortcomings; the objects being measured and standards lack in intensity of blackness.

In order to increase the reliability of bearing assemblies and optimize the thermal condition of the working zone of a bearing, the method of electromodelling of temperature pattern, including the working surfaces of bearings, was suggested. This method is based on the principle of electrothermal analogy, i.e., on mathematical description of distribution of thermal flows in the unit under study and electric current on a model made from electrical conductive paper [8]. An electrical model of an axially symmetric component is made in a shape similar to the corresponding zones in an actual unit. If there is an axial symmetry, the thermal conductivity is described by Laplace equation in cylindrical coordinates

$$\partial^2 u/(\partial r^2) + (1/r)\,\partial u/(\partial r) - \partial^2 u/(\partial r^2) = 0$$

Then, by analogy, the conductance of the medium along the axis must be constant, and along the axis z it changes according to the linear law, i.e.,

$$\sigma z = \text{const}; \quad \sigma r = kr$$

The paper with uniform conductance is used for convenience, and the conductance of zones of interest are being changed by the use of paper sheets bonded by an electroconductive adhesive.

Thermal resistance on the mounting fits and in elements of the unit may be simulated by holes on the boundary between the elements. The ratio of pitch of the holes to a width of paper between the neighboring holes must be equal to the ratio of thermal resistance of metal to thermal resistance of the joint. Depending on the character of heat exchange, it is possible to obtain data on the thermal heat resistance of the joint experimentally or from reference publications.

The analytical survey made it possible to draw a conclusion that temperature measurement on surfaces of solid bodies by means of thermocouples is rather widely used.

Thermocouples, however, develop a very small thermoelectromotive force, and due to that, very sensitive d.c. amplifiers must be used.

The method of the measurement of temperature by metallic thermoresistors has a linear thermoelectric characteristic but it is inconvenient for balancing errors due to a change of resistance of connecting wires during their aging and because of changes in temperature of the surrounding medium.

Semiconductor temperature-sensitive resistors have rather a high sensitivity when compared to thermocouples. They are able to work at low temperatures but dimensions are rather large.

The optical method is used for the contactless control, but in measuring the temperature of small surfaces the measurement is not sufficiently accurate.

The method of electromodelling allows to solve a number of new engineering problems, such as evaluation of: Integral temperature of a lubricating layer in the working zone without dismantling the bearing; thermal resistance of a bearing; radiation of units with several heat sources and sinks; and the evaluation of a heat flow from the bearing to a housing.

3.6 DEVICES FOR DIAGNOSTICS OF ROLLING CONTACT BEARINGS

Diagnostic devices for bearings are designed for the analysis of their technical condition. The technical condition of bearings is determined by errors due to manufacturing and assembly processes.

In terms of technical diagnostics it is assumed that an object may be in an infinite set of states, which in their turn may be divided into two subsets, N_1 and N_2. As a rule, the transition of an object from one state to another is explained by a fault of some element occurring in the object. The possible faults are divided into the faults of an element of the object itself.

The subset N_1 includes each state in which the object can perform all the operations or solve the tasks assigned to it, i.e., the operational states. Each of the states of this subset differs in degree or margin of efficient operation, which is characterized by the distance of the state from the limiting state. Transition from one state to another in the subset N_1 may be explained by appearance of faults in the object which do not result in the loss of operation ability, i.e., do not cause the transition of the object due to its state into the subset N_2. The subset N_2 includes all the states in which faults in the object result in the loss of capacity for work. The size of the subset N_2 is determined by the number of faults which may be detected from the respective indications. Such classification of the state of an object makes it possible to divide the process of technical diagnostics into two stages.

In the first stage the attribution of an object to one of the subsets N_1 or N_2 is established. In a number of instances it is necessary to determine what kind of change of state takes place in the case under consideration. This stage may be called the determination of operation ability. Analysis of the states of an object in the subset N_1 allows determining the degree of a change in its capacity for work, and in a number of cases it helps to forecast the moment of transition of the object into subset N_2. Hence, the forecasting of the technical state of an object has been accomplished. The success of diagnostics is largely determined by the knowledge of working conditions and the possible means of monitoring indices characterizing the changes of the state of an object in time.

At the second stage it is established at which subset the state of the object belongs to the subset N_2. This stage may be called the detection of a developing fault. Data on temperature, the state of lubricant, power losses (friction torque), as well as vibroacoustical characteristics may be used as indicating signals on the state of the object. Merits and demerits of each of these diagnostic parameters are presented in the literature [21, 25, 38].

In works on technical diagnostics the methods of vibroacoustical diagnosis (forecasting) take the leading position. In vibroacoustical diagnostics, natural vibrations or the noise of a bearing are considered as indicating signals. Vibration diagnostics is a nondestructive method of evaluation. The advantage of such methods lies in the fact that practically all forms of faults—manufacturing, mounting, and operational—can be diagnosed with the help of these methods.

All acoustic methods are subdivided into active and passive. The active methods are based on excitation and reception of acoustic and ultrasonic waves, which is done with the help of specialized transducers. The state of the product under control is judged by the parameters of elastic waves passed through the product. This method is of little use for the control of rolling contact bearings since elastic waves are reflected from the inner structural cavities as well as from the inner defects.

The passive methods are subdivided into noise/vibration and acoustic emission. All passive methods are based on recording and analysis of elastic waves arising in the products themselves. During the application of the noise/vibration methods, the noise and vibrations arising during the work of a bearing are used. Here, different parameters of a signal can be used. However, for the purpose of diagnostics the analysis of the spectrum of oscillations and the study of laws of distribution of vibration amplitudes have gained the widest use.

The method of acoustic emission is based on recording of elastic waves arising at the moment of formation and development of cracks. In this case the fault in itself is the source of ultrasonic waves. Thus, the method of acoustic emission, in contrast to the noise/vibration method, allows forecasting the appearance of inner defects.

The characteristics of the equipment which implements the passive control methods are determined by properties of acoustic radiation. Such apparatus must be constituted from the following units: primary transducer (most often piezoelectric); low noise amplifier; signal processing system; recording unit. The major difficulties in the development of such apparatus are due to the fact that it is necessary to receive extremely weak signs of emission (oscillatory displacement of 10^{-14}–10^{-7} m) at the noisy background, and the equipment must have very fast response, the minimum length of pulses reaching 10 ns. The development of wide band primary transducers is a task of particular complexity.

The signal processing system is determined by the chosen method of measurement. It must include the analyzers of spectrum and of amplitude distribution of the received signals. In recent years, the so-called analyzers of cepstrum for treatment of such signals have come into use. In the general case, the following relationship is called the complex cepstrum:

$$C_l(\tau) = F^{-1}\{\lg Fx(f)\}$$

where $Fx(f) = f\{fx(t)\}$ is the complex spectrum of the oscillation; $fx(t)$ is the direct Fourier transform; F^{-1} is the inverse Fourier transform.

The cepstrum is used for diagnosing the technical state of turbines and gear boxes and it allows to detect the faulty rotating elements. The cepstrum may also find application for diagnosing rotating bearings. Standard devices for acquiring the cepstrum are not yet available in the USSR.

The general block diagram of the setup for diagnosing technological errors during manufacture and assembly of bearings is presented in Fig. 3.16 [52]. The vibrotransducer turns the mechanical vibration of a bearing under test into electric signal. The amplifier-corrector provides the required frequency characteristic of the measuring channel. The spectrum analyzer carries out the measurement of amplitudes of vibrations of a bearing at different frequencies. As a rule, the analyzers of sequential operation are used in devices for diagnosis of bearings operating in steady regimes. Analyzers of parallel action are used for diagnosis of bearings working in quasisteady regimes, i.e., with variable rotation speed. They include filters of informative frequencies connected in parallel. The application of analyzers of parallel action allows an appreciable reduction of diagnostic time, though it reduces the universality of the system.

The arithmetic unit is designed for computation of the value of the decision function. The characteristic or parameter being diagnosed is recorded from the value of decision function in the indication unit.

The block diagram of the device for detection of the character of technological errors during the processes of manufacturing or assembly of bearings

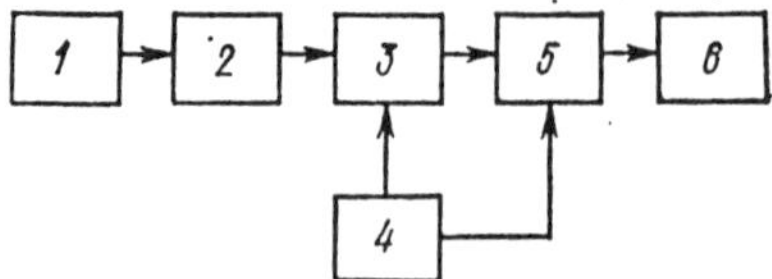

Figure 3.16 General block diagram of diagnostics setup: *1*) vibrotransducer; *2*) amplifier-corrector; *3*) spectrum analyzer; *4*) control unit; *5*) arithmetic unit; *6*) indication unit.

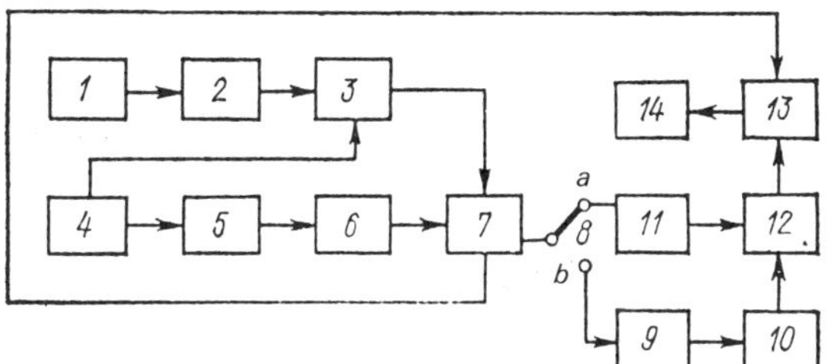

Figure 3.17 Block diagram of device for determining errors in manufacture and assembly of bearings for steady regime operation: *1*) vibrotransducer; *2*) amplifier-corrector; *3*) spectrum analyzer; *4*) synchronizer; *5*) binary counter; *6*) matrix decoder; *7*) coincidence circuit; *8*) regime switch; *9*) computation circuit of weighting coefficients; *10*) circuit of inputting weighting coefficients; *11*) memory circuit; *12*) summator; *13*) threshold circuit; *14*) indication unit.

operating in a steady regime (Fig. 3.17) includes the control unit, which comprises a synchronizer, binary counter, and matrix decoder. The arithmetic unit incorporates a coincidence circuit, the regime switch, computational scheme of mass coefficients, input circuit for mass coefficients, memory circuit, summator, and threshold circuit.

In this setup a spectrum analyzer of sequential action is used. Information on the state of the counter comes to the matrix decoder, the codes of information frequencies being displayed on its commutation panel. The signal from the matrix decoder reaches the coincidence circuit only if the spectrum analyzer is tuned to one of the informative frequencies. To the second input of the coincidence circuit the signal from the spectrum analyzer is supplied which is proportional to the amplitude of vibration on the tuned-in frequency. The coincidence circuit carries out the counting of the values of functions $e(x_{jp} - \Delta x_{jp})$ on each informative frequency and the memory circuit does the recording. The summator is calculating the value of discriminative function $g(x_j)$. The threshold circuit determines the sign of $g(x_j)$. The indication (display) unit indicates occurrence of a technological defect of one or other sort in the test bearing. The arithmetic unit operates in two regimes, recognition and learning. In the recognition regime, the switch *8* is connected to the terminal *a*, while in the learning regime it is connected to the terminal *b*. For testing of the diagnostic device, vibration characteristics of bearings with known technological errors are used. In the learning regime the computational circuit for coefficients carries out the following transforms with the output parameters P_p, Q_p and P_1:

$$V_p = \lg \{P_p Q_p/[(1 - P_p)(1 - Q_p)]\}, \quad p = 1, 2, \ldots, d$$

$$V_{d+1} = \sum_{p=1}^{d} \lg [(1 - Q_p)/P_p] + \lg [P_1/(1 - P_1)]$$

and the input circuit of mass coefficients performs their setting into the summator. This device determines the type of technological defect of the bearing:

ovality and poligonality of inner and outer races, and dimensional scatter of rolling elements.

In many diagnostic devices, piezoelectric elements are used as transducers (Fig. 3.18).

Various techniques and devices for implementing the method of acoustic emission are based on application of forced sources of vibrations. For example, with the help of an electromagnetic vibrator oscillations in a product are excited. These oscillations are received using a transducer. Then with the help of a measuring instrument, the amplitudes of the principal and additional resonances are recorded. In the case of quality control of a suitable product the parameters of the basic resonance do not depend on the position of the bearing ring with respect to the point of suspension, and the additional resonance is absent. During the control of a product with a defect, both resonances, the basic and additional, are present. The ratio of the amplitudes of resonances varies with the change of the position of the defect in respect to the suspension point of the product. When the ratio of amplitudes of the additional and principal resonances is zero, the defect is in the cross section of the bearing ring at which it is suspended.

Analysis of the available vibration diagnostics equipment enables us to draw the following conclusions.

For the assessment of defects or diagnostics of objects with rotating elements on rolling contact bearings the most general method for analysis is the observation over the changes of mean square level and spectral power of a vibroacoustical signal. In this case, however, the parameters depend on loading on the bearing, rotational speed, the tightness of mounting of a bearing, the amount of lubricant, etc.

During the measurement of probability characteristics of a vibroacoustic signal, such as probability density, mathematical expectation, dispersion, excess, etc., the probabilistic moments supply the information on the state of an object regardless of rotation speed or loading on the bearing.

In existing instruments and devices, as well as in ones under development, for diagnosis of the technical state of ball bearings in an assembled object or in self-contained condition the vibroacoustic signal is the most recognized diagnostic parameter.

Diagnostics usually is carried out by means of a comparison of amplitudes of the spectrum of a vibroacoustic test signal with the spectrum of the standard bearing or with the assumed threshold value of theoretical spectrum. The "contradiction" with respect to optimum regions of frequencies for diagnosis of bearing assemblies is quite difficult to resolve.

Discrete pulses commonly related to defects arise in different frequency ranges with various degrees of conspicuousness. This distinctive feature may be explained in qualitative terms by considering the relative levels of the disturbance signal caused by the damage and background noise. The efficiency in detecting a fault is closely dependent on the level of background noise. The

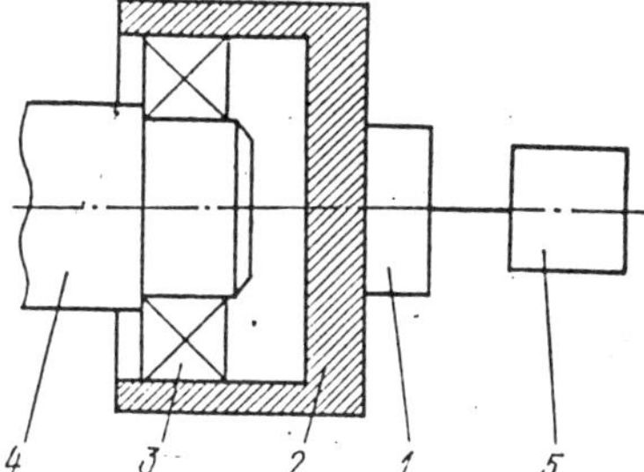

Figure 3.18 Block diagram for estimating technical condition of rolling contact bearings: *1*) vibration receiver; *2*) housing for test bearing; *3*) test bearing; *4*) drive shaft; *5*) recorder.

principal difficulties connected with the development of vibroacoustic equipment for the purposes of diagnostics lie in the fact that it is necessary to receive very weak signals and the apparatus must be very responsive. In addition to the vibroacoustic signal, the ohmic and inductive impedance are recognized as diagnostic parameters. The change in lubricating oil film is used in some diagnostic instruments and devices.

Transient resistance in bearings depends on rotational speed and loading on the bearing, also on the presence of wear products in lubricating oil. Therefore this diagnostic parameter is of an integral nature. For localization of a fault according to the resistance it is necessary to have the data on the working regime of the bearing during the diagnosis (rotation speed, loading) and information concerning the presence of conductive products in lubricating oil.

As it is shown in [40, USSR certificate of invention 540186], the noise of bearings has direct correlation with the axial and radial vibrations of a bearing. The methods of quality control of a bearing based on the measurement of noise may supply us with additional useful information especially in the range of 500–700 Hz. In some countries (for example, in Japan) standard quality control methods for bearings based on the measurement of noise are universally adopted. The choice of equipment for measuring noise of instrument bearings depends on the measurement conditions, the way of recording the obtained results, etc. In the majority of cases such equipment consists of the following principal units: small-size cabinets, measuring microphones, sound level meters, frequency analyzers, and recorders.

As computations and experimental studies show, instrument bearings produce noise within a wide frequency range (20 Hz–20 kHz). The measurements of bearings are carried out in specialized chambers (cabinets) with the following specifications: the frequency range of the noise measurement is 20 Hz–20 kHz; the dynamic range of the noise level measuring is to be up to 50 db; the setup is not to render any effect on measurement of noise; the allowable levels of the background noise in the range of frequencies under measurement should be at least 10 db lower than the level of noise from bearings being tested.

The diagnostics equipment also may include the forecasting apparatus, which is meant for evaluation of technical state and parametric resource of bearings. In the forecasting process, the information on changes of bearing parameters related to their technical condition is used. A general block diagram of the forecasting setup [39] is shown in Fig. 3.19. In the setups under construction, the natural vibrations of bearings are used as parameters N according to which the forecasting is made.

The extrapolator carries out the simulation of the law of parameter change in time taking into account the initial value N_0, which is presented in the form of function $\varphi(k_1, N_0)$. The control unit periodically supplies the extrapolator and arrangement of display with the signals on preparation and commencement of simulation of the change of the parameter N. During this process, calibration of the simulated processes of the change of technical state is performed. The threshold unit compares the current value of the parameter of forecasting with its critical value and at the moment of their coincidence there develops a signal of the ending of simulation which is supplied to the display unit. Thus, the display unit periodically records the parametric working life of the bearing, which is obtained as a result of simulation of solution of the equation (Fig. 3.20)

$$\varphi(t, N_0) = N_{cr}$$

The forecastor may be used in forecasting for bearings working in steady or quasisteady regimes. It can forecast failures of the bearing. The forecasting is carried out as a result of an analysis of a change of the forecasting parameter within a certain period of time t^* and its comparison with the critical change $J_{cr}t^*$, where J_{cr} is the critical intensity of change for the forecasting parameter.

The process of simulation from which the state of failure is evaluated is determined by

$$Jt^* = k_1 g^* t^* + k_2 g^{*\prime} (t^*)^2$$

where J is the intensity of change of the forecasting parameter of the bearing under test.

If $Jt^* < J_{cr}t^*$, then the bearing is regarded as functioning normally. If $Jt^* \geq J_{cr}t^*$, then it is possible to expect failure of a bearing in the near future. The value of critical intensity of change of the forecasting parameter is determined on the basis of the resource tests of the bearing or simulation of change of its technical condition on computers.

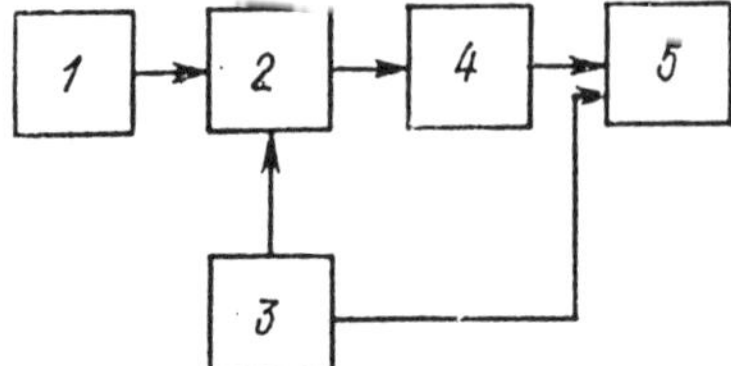

Figure 3.19 General block diagram of the forecasting system: *1*) diagnostics arrangement; *2*) extrapolator; *3*) monitoring unit; *4*) threshold unit; *5*) indication unit.

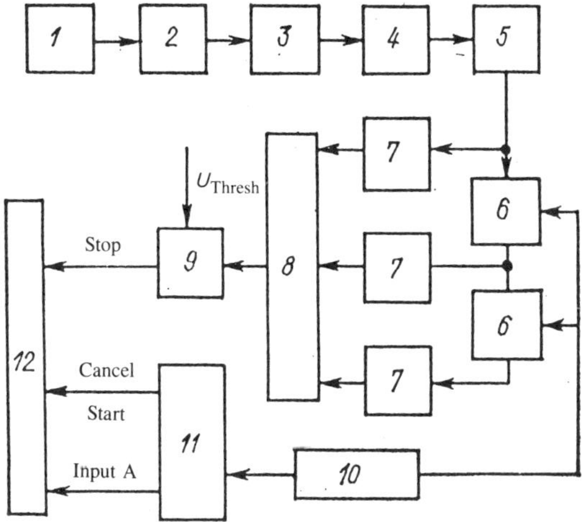

Figure 3.20 Block diagram for forecasting device for parametric service life of bearings: *1*) vibrotransducer; *2*) amplifier-corrector; *3*) spectrum analyzer; *4*) arithmetic unit; *5*) digital-to-analog converter; *6*) integrators; *7*) weighing circuits; *8*) summator; *9*) threshold circuit; *10*) master oscillator; *11*) display control unit; *12*) indication (display) unit.

3.7 DYNAMIC AND ACCURACY CHARACTERISTICS OF MEASUREMENT MEANS OF RESISTANCE TORQUE

The process of measuring any physical quantity is an operation whose results show by how much the quantity being measured differs from the reference value. No single measurement can be carried out with absolute accuracy. When measurements are carried out with the help of measuring instruments errors are committed which are determined by the errors in the measuring instrument itself. The problem includes not only determination of the measured quantity but also an evaluation of the error involved in the measurement.

For the diagnostic systems and instruments under consideration the major errors will result from several sources: the dynamic errors in the measuring circuitry of the instruments; the errors of sensors which can be attributed to systemic errors; random errors due to non-considered changes in operational conditions; and factors of the environment.

The dynamic error of measuring circuits is determined as a difference between the actual quantity being measured and the indicated quantity: $\Delta x = x_{act} - x$. The dynamic error in percentage shows the relative error $\Delta x = (\Delta x/x)100$.

In estimating dynamic errors of instruments it is very important to match the testing condition with the dynamic qualities of measuring instruments. In determining the friction torques the frequency range of quantities being measured and the sensitivity of the setup are of major importance. They are interrelated quantities. In order to increase the frequency range it is necessary to raise the natural frequency of the system as high as possible. When computing the natural frequency in the instruments considered here, the most important part of the moving system is the moving parts of the test bearing or the rotational system. Hence it is necessary to accurately compute the moment of inertia of the bearing. Let us consider here as an example the moment of inertia of a bearing. The kinetic energy of a rolling contact bearing is equal to the sum of the kinetic energy of its parts:

$$E = E_2 + E_3 + E_4$$

where E is the kinetic energy of the moving parts of the bearing; E_2 is the kinetic energy of the moving rings of the bearing; E_3 is the kinetic energy of all rolling elements; E_4 is the kinetic energy of the cage.

To compute the moments of inertia it is necessary to consider the bearings with moving outer and inner rings separately (Fig. 3.21).

The formulas developed below are suitable for the computation of moments of inertia of all types of rolling contact bearings except thrust bearings, for which the concept of outer ring is not applicable.

For a bearing with stationary outer ring

$$E = J\omega_{21}/2; \quad E_2 = J_2\omega_{21}^2/2; \quad E_3 = n\left(J_3\omega_{31}^2/2 + m_3 v_{\text{cen31}}^2/2\right)$$
$$E_4 = J_4\omega_{41}^2/2$$

where J is equivalent moment of inertia of the bearing; J_2 is the moment of inertia of the moving ring *2*; J_3 is the moment of inertia of the rolling elements *3*; J_4 is the moment of inertia of the cage *4*; ω_{21} is the angular speed of the moving ring *2* relative to the stationary ring *1*; ω_{31} is the angular speed of the rolling elements *3* relative to the stationary ring *1*; ω_{41} is the angular speed of the cage *4* relative to the stationary ring *1*; m_3 is the mass of rolling elements *3*; v_{cen31} is the velocity of the center of gravity of rolling elements relative to the stationary ring *1*.

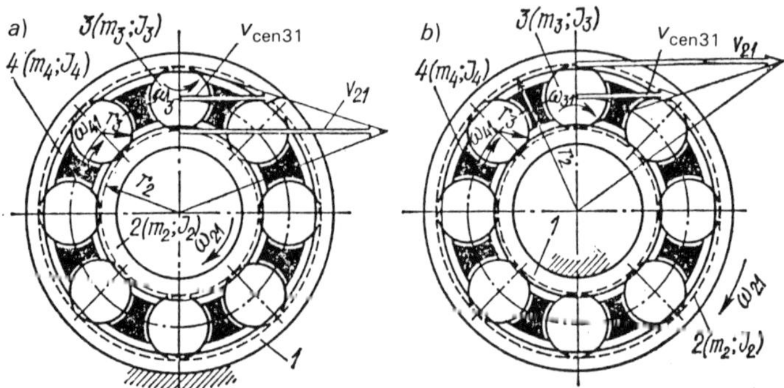

Figure 3.21 Computation of moments of inertia of rolling contact bearings: *a*) moving inner ring of bearings; *b*) moving outer ring of bearing.

After necessary transformations of the above expressions, the moment of inertia of a bearing with a stationary outer ring will be given by

$$J = J_2 + n/4\,[J_3(r_2/r_3)^2 + m_3r_3^2] + J_4/4\,[r_2/(r_2 + r_3)]^2$$

Accordingly, the moment of inertia of a bearing with a stationary inner ring is

$$J = J_2 + n/4\,[J_3(r_2/r_3)^2 + m_3r_2^2] + J_4/4\,[r_2/(r_2 - r_3)]^2$$

The moments of inertia of the remaining elements of the measuring systems are computed from the well-known formulas.

In considering the natural frequency of the system it is necessary to keep in mind that it should be 2–3 times larger than the upper bound of the operating frequency of the process being studied, in this case, friction torque.

Knowing the equivalent moments of inertia of the measuring system one can generally assume that the mass of the moving part of the instrument and the device $m = J/R^2$, where R is the distance from the center of rotation to the strain gauge which senses the torque being measured. Additional inertial forces F would act on the strain gauges in case of vibrations or runout and would cause an error:

$$F = ma$$

where m is the equivalent mass; a is acceleration.

Then the error in the measurement of the variable component of friction torque of the bearing will be (%)

$$\Delta F = (F/F_{var})100$$

where $F_{var} = M_{var}/R$ is the force acting on the system of gauges while measuring the variable component of the friction torque. Here R is the distance from the bearing center to the gauge of the measuring system.

Another source of error in the measurement of friction torque components is the nonuniformity of rotation of the driving unit. The principal sources of nonuniformity of rotation are the following: *a*) nonuniform speed of electric motor transmitted to the bearing through the drive; *b*) radial runout of pulleys (if any) which directly influences the rotational speed of the bearing and also adversely affects the working of the system by exciting longitudinal vibrations in the belt, etc.; *c*) unbalanced masses; *d*) the rolling contact bearings of the drive, etc.

Tests at Kaunas Polytechnic Institute for nonuniformity of synchronous motors of the DSG type showed that their nonuniformity of rotation decreases with increased speed (Table 3.4).

The tests also showed that in belt drives an increase in the tension of the belt increases the nonuniformity of their rotational speed. This is explained by the fact that with the increased tension the stiffness of the belt increases, i.e., the natural frequency of the belt also changes. In the resonance region this leads to an increase in the coefficient of nonuniformity. Hence determination of the coefficient of nonuniformity of rotation of the moving ring of the bearing is necessary in the study of errors in measurement. For this purpose it is possible to use different devices (for example, a trigger marker of the 51E10 DISA type which gives a signal after a specified angle of rotation).

The methodology for determination of the coefficient of nonuniformity is as follows: at first we determine the frequency of flow fluctuation which depends on the rotational speed of the marker disc, $f = ns$, where s is the number of pulses per revolution and depends on the angle set for producing a signal; n is the rotational speed of the moving ring (s^{-1}), which is selected depending on the operational regime.

For convenience of computation and increased accuracy both signals can be recorded on photo film by high-speed photography.

The coefficient of nonuniformity in rotational speed is determined as

$$\delta = l/(Lf\,\Delta t)$$

where l is the maximum phase shift; L is the period of reference signal; f is the pulse frequency; Δt is the time for phase shift.

Measures for reduction of errors in measuring resistance torque. All principal errors may be classified in the following way:

1) dynamic errors are resonance- and inertia-related errors resulting from vibration of a bearing assembly with the test bearings relative to transducers and vibration of the base on which the transducers are installed;

2) errors due to inadequate directionality of transducers;

3) errors as a result of disturbance of the test unit;

4) errors at amplification and processing of an electric signal;

5) errors due to nonuniformity of rotation of the drive.

All these groups of errors may be rather large, sometimes exceeding the signal being measured.

In order to understand measures used to alleviate dynamic errors we will consider dynamics of the measuring system in Fig. 3.22. It works as follows: during the rotation of the drive shaft *1* on which the test bearing *2* is mounted, the friction torque through the rod *3* and the hard stone supports with the needle

TABLE 3.4 Nonuniformity of rotation of motors of the DSG type

Type of electric motor	Speed, rps	Nonuniformity of rotation, %
DSG-1	25	0.16
	50	0.06
DSG-2	25	0.40
	50	0.15

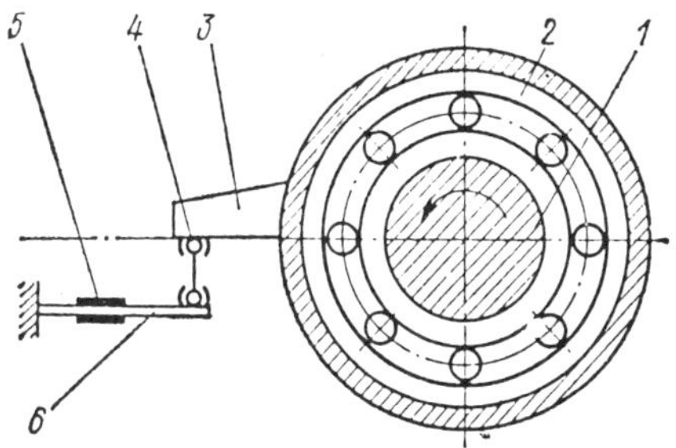

Figure 3.22 Classical schematic of torque meter.

4 is transmitted to the elastic small beam to which the strain gauges *5* are bonded.

The test system in the general case is a system with three degrees of freedom (Fig. 3.23). Here K_1, K_2, K_3, K_4 are elasticities of the rod *3*, needle *4*, elastic beam *6* (Fig. 3.22), and of the contacts in the bearing *2* between the balls and races, respectively; C_1, C_2, C_3, C_4 are the coefficients of damping of the rod *3*, contacts of the needle *4*, attachment of the elastic beam *6*, and the bearing *2*, respectively; J_1, J_2, J_3 are the moments of inertia of the mobile part of the test assembly, needle *4* and elastic beam *6*, respectively.

The moments of inertia J_2 and J_3, as compared to the moment being measured, are negligible and in computations may be neglected. Thus the dynamic model is simplified (Fig. 3.24). Then

$$K = K_1K_2K_3/(K_2K_3 + K_1K_3 + K_1K_2) + K_4$$
$$C = C_4\,(C_1 + C_2 + C_3)/(C_1 + C_2 + C_3 + C_4)$$

Accordingly, the differential equation of free motion becomes

$$J\ddot{\varphi} + c\ddot{\varphi} + k\varphi = 0$$

where $\ddot{\varphi}$ is the angular acceleration of the moving part of the system.

The natural frequency of torsional vibrations with damping is expressed as follows:

$$\omega_y = \sqrt{1 - \xi^2}\,\omega_c$$

where $\xi = C/(2J\omega_c)$ is the damping ratio, $\omega_c = \sqrt{K/J}$ is the natural frequency of torsional vibrations without damping in expanded form:

$$\omega_c = \sqrt{[K_1K_2K_3 + (K_1K_2 + K_1K_3 + K_2K_3)]/[J\,(K_1K_2 + K_1K_3 + K_2K_3)]}$$

Finally,

$$\omega_y = 1/2J\sqrt{4J\,[K_1K_2K_3 + K_4\,(K_1K_2 + K_1K_3 + K_2K_3)]/(K_1K_2 + K_1K_3 + K_2K_3) - [C_4\,(C_1 + C_2 + C_3)/(C_1 + C_2 + C_3 + C_4)]^2}$$

The reduced moment of inertia is

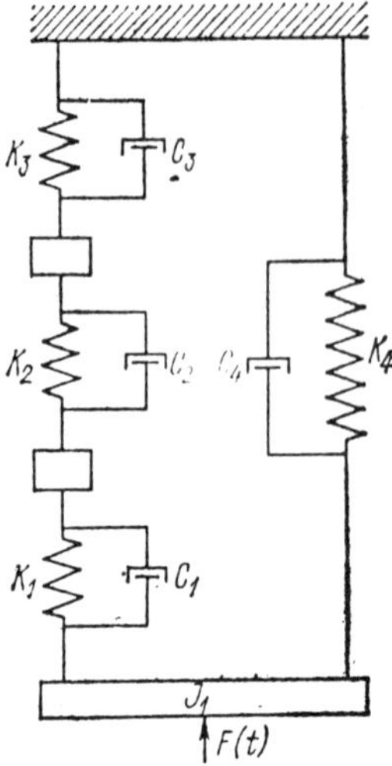

Figure 3.23 Dynamic model of measuring system.

$$J = J_1 + i^2 J_2$$

where J_1, J_2 are the moments of inertia of the moving part of the test unit; of the disc, and of the moving part of the bearing; i is the transmission ratio.

Practically, in instruments under consideration, the systematic and random errors of measurement are close to one another. The accuracy of the results are determined by them to the same extent, hence it is possible to use a total error. If the magnitude of systematic error is designated as δ, and the dispersion of measurement results is σ^2, then the upper boundary of total error is $\Sigma = \delta + 2\sigma$.

This rule of addition (with probability higher than 0.95) may be applied to a great variety of measurements. Addition in the general case of systematic and accidental errors is possible only in the case when one of the errors is no more than one order of magnitude greater than the other.

Calibration of the torque meters. An increase of measuring accuracy requires continuous perfection of calibrating devices for measuring instruments. Calibration, in its turn, depends on the methodology chosen and on instruments used during the calibration process [7]. The greatest difficulties arise at the calibration of microdynamometric devices possessing high sensitivity (to

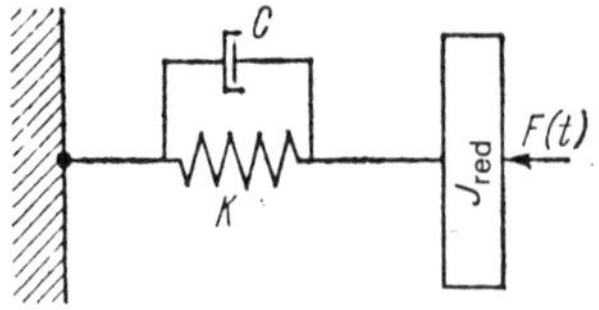

Figure 3.24 Simplified dynamic model of measuring system.

$1 \cdot 10^{-3}$ Nm and higher) which are sensitive to the action of convective flows of air, changes in humidity, temperature, magnetic disturbances, etc.

Direction of the sensitive axis is a significant design feature in calibration. Specifically, the horizontal arrangement of the axis allows carrying out calibration by force of gravity.

While carrying out the dynamic calibration for dynamometric equipment working in conditions of variable loading it should be borne in mind that first of all the amplitude-frequency characteristic should be measured. If the amplitude-frequency characteristic is constant, $x = A$ in the whole frequency range, then the calibration may be carried out in statics.

Calibration of microtorquemeters may be carried out by the action of: force of gravity; restoring forces of elastic elements, or forces and torques of electrical, magnetic, and electromagnetic nature; buoyancy Archimedean force; force of light pressure (range of ultrasmall moments).

We will consider here a few techniques for graduation of torquemeters of high accuracy used in dynamometry.

In the weight-type torquemeters with horizontal axis of sensitivity, it is more advisable to use the technique of weight—arm for calibration. The standard weights are used, whose mass is known with high accuracy. With this technique the main sources of error are as follows: *a*) errors due to the class of accuracy of the analytical balance and of a set of weights; *b*) errors due to the presence of zones of insensitivity in a measuring null-indicator and the friction torque in knife-edge support; *c*) errors in determining linear quantities.

Analysis shows that in using analytical scale model VLM-20 (with the error up to ± 0.093 mg) and tool microscope model BMI-1 (with the error up to ± 3 to 5 μm) the relative error of calibration does not exceed 0.02–0.04% of the full-scale value.

The use of the torque reproduction technique with the help of elastic suspensions, yields the error which is accumulated from the following component errors: duration of oscillation periods of suspension system; the moment of inertia of calibration element; torsion angle of suspension; presence of lateral component at the noncentral attaching of suspension; instability of elastic qualities of suspension due to temperature effects.

The analysis shows that with metallic suspensions of high stability, as well as suspensions and filaments of quartz, the relative error of calibration does not exceed 0.05–0.1% of the full-scale value.

In the majority of calibration cases the high-stability electric torque transducers can be used. Under stabilization of temperature the class of accuracy of such instruments may be 0.01–0.04 and even lower. In this case calibration is carried out by the torque transfer method.

At the first stage the torque transducer is installed into the standard torquemeter whose sensitivity is known; a necessary electric signal (for example, current) $I_{\text{cal}1}$ is supplied to the transducer; the compensation (measurement) of electromechanical moment created by the transducer is performed. Then the torque constant of the transducer (slope of torque-current characteristic) is determined as

$$K = T_K/I_{cal1} = T_{cal1}/I_{cal1}$$

where $T_K = T_{cal\,i}$ are the compensating and electromechanical torques.

If needed, the corrections on nonlinearity of the function $T_{av} = f(I_{cal})$ are determined in a similar way.

At the second stage, the required quantities of currents $I_{cai\,i}$ corresponding to reference points of the dial are computed; usually the ratio $T_{cal\,i}/T_{cal1} < 1$, $I_{cal\,i} = (T_{cal\,i}/T_{cal1})I_{cal1}$ is used.

At the third stage, the torque transducer is transferred to the torquemeter being calibrated. Current I_{cal} in the transducer are measured and the meaning of one division in different segments of the dial scale are determined.

The basic sources of errors using this calibration technique are: error of reference torque transducer ΔT_{ref}; measurement errors of $\Delta I_{cal\,i}$, due to instability of transducer constant in time, $\Delta T_{cal}/I_{cal}$; error due to the inaccuracy of adjustment of the torque transducer $\Delta T_{cal\,dev}$.

Dispersion of calibrated moment

$$D(T_{cal}) = D(\Delta T_{ref}) + (T_{cal}/I_{cal})^2 D(I_{cal}) + 2I_{cal}^2 D(T_{cal}/I_{cal}) + D(\Delta T_{cal\,dev})$$

Analysis shows that the cumulative error of calibration according to the transfer technique with application of magnetic electrical torque transducers and reference standard torquemeters is 0.02–0.08%. The advantages of portable electrical torque transducers lie in the fact that they increase the mobility and rapidity of calibration as well as reduce mechanical effects on test instruments.

CHAPTER

FOUR

ANALYTICAL STUDY METHODS OF BEARING CHARACTERISTICS

The techniques for study of bearing characteristics are based on the application of diagnostic symptoms according to which the judgement of the work capability of a bearing is made. All symptoms used for the dynamic diagnosis are utilized in analysis of vibratory processes taking place in bearings and their assemblies. Now it is possible in terms of mathematics to formulate stringent methods of computation of basic statistical characteristics of peak values of random processes which develop due to deviations from the true geometric shape. These random processes result in generation of variable component of the friction torque in vibrations of bearing elements or the complete assemblies, in changes of hydrodynamic lubricating film, and in other diagnostic symptoms.

The method of acoustic emission becomes gradually more and more popular. This method is based on the fact that the approaching fatigue failure is always followed by pitting or separation of steel either on race surfaces or on balls or rollers. When the balls or rollers roll over the damaged place mechanical kicks create brief transitory high-frequency oscillations which are transmitted to the housing of a bearing. Regular mechanical vibrations do not exceed 30 kHz, and impact waves arising in a faulty bearing frequency are reaching 50 kHz and more.

Monitoring of impact pulses is based on determination and decoding of probabilistic characteristics of random processes.

4.1 PROBABILISTIC ANALYSIS

A bearing or rotor system as an object of diagnostics may be considered in two aspects, namely, from the points of view of structure and of the functioning condition. Each of these aspects has its specifics.

In the structural approach we deal with dimensions and shapes of bearing elements, with clearances in kinematic pairs, and other properties ensuring the normal operation of a bearing. The condition of a mechanism or system is the basic concept of the structural aspect of diagnostics.

From the point of view of functioning, a bearing is looked upon as a uniform system generating different processes: oscillations, radiation of heat, etc.

At a certain point of time t the structural properties of a bearing and its operational processes may be described by sets of parameters. It is assumed that an ideal bearing whose structure exactly corresponds to the original design has the reference values of parameters. In comparison with these parameters it is possible to judge the state of a bearing.

The parameters describing the structure of a mechanism are denoted by x_i, $i = 1, 2, \ldots, n$, and the parameters which will describe the functioning or a condition of a bearing we denote by y_j, $j = 1, 2, \ldots, m$. The initial (reference) values of the parameters we denote by x_{i0}, $i = 1, 2, \ldots, n$ and y_{j0}, $j = 1, 2, \ldots, m$, measured values, i.e., the values of the parameters at the moment t, are x_{it}, $i = 1, 2, \ldots, n$ and y_{jt}, $j = 1, 2, \ldots, m$.

Difference $y_i = x_{it} - x_{i0}$, $i = 1, 2, \ldots, n$ describes deviation of a certain parameter of bearing being diagnosed from the parameters of its ideal prototype.

On the set of structural parameters x_i, $i = 1, 2, \ldots, n$ the minimality condition is imposed. It is required that none of the values x_i, $i = 1, 2, \ldots, n$ from the point of view of function be expressed through the values of other parameters included into the set. For example, since such parameters of a bearing as the diameter of a shaft d, the diameter of a bushing D, and clearance b between them are linked with the relationship $b = D - d$, one of them is excessive. On the other hand, the set of parameters must be complete because the parameters enable making a unique decision on the state of the bearing.

To the set of functional parameters we attribute also the functions of state, which include the numeric characteristics of processes accompanying the operation of bearings. These are, first of all, measurements of total lubricating film, friction torque, and vibrations. Along with them the state of lubricant, temperature condition of bearing races, and variable environmental conditions have to be analyzed. All these parameters are accessible to direct measurement at a certain moment of time or even at some period of time. They serve as sources of information on the state of a bearing.

The set of functional or state parameters may also be required to comply with independence conditions (or minimality condition) and completeness. It should be noted that due to inescapable errors of measurement of actual parameters of both signals and processes, the set of respective parameters should comply with a somewhat weakened requirement of completeness. In other words, in order to increase the accuracy and reliability of diagnostics it is necessary to use the redundant information.

Thus for the description of the state of a bearing we have two sets of parameters: $x_1, x_2, \ldots, x_n$ and $y_1, y_2, \ldots, y_m$. In order to solve a diagnostic problem it is necessary to have one-to-one correspondence between the two sets of parameters, i.e., for each set of values $x_1, x_2, \ldots, x_n$ (i.e., for each possible state of a bearing), a system with the values $y_1, y_2, \ldots, y_m$ must correspond, and vice versa.

The very important and the most time consuming stage of development of a diagnostic methodology is determination of dependences of the state parameters in the measured functional parameters $x_i = y_i(y_1, y_2, \ldots, y_m, i = 1, 2, \ldots, n$.

Here the method of dispersion analysis may be used. With its help we evaluate the influence of a certain parameter (for example, the clearance between the race and the ball) on the change of other state parameters of a bearing. Further, we will assume that random measurement errors have a normal distribution. As a result of measurement of individual parameters, the actual (mean) value a and dispersion σ^2 of the other parameters may vary. At the start we will assume that the dispersion remains invariant. If stability of dispersion evokes doubts it is necessary to carry out additional investigation using criteria of Bartlet or Kokhran number.

If the analyzed parameter x_i at different values leads to a series of results x_i, $x_{i2}, \ldots, x_{ik}$, then as an indicator of the influence of the parameter x we will take the number

$$\sigma^2_{xi} = (1/k) \sum_{i=1}^{k} (x_{il} - \bar{y}_j)^2$$

where $\bar{y}_j$ is the arithmetic mean of the numbers y_{j2}; then $\bar{x}_i = (1/k) \Sigma^k_{i=1} x_{il}$.

The number σ^2_{xi}, analogously with an ordinary dispersion, is called dispersion of the parameter $\bar{x}_i$, though the numbers $x_1, x_2, \ldots x_{in}$ are not random ones; σ^2_{xi} is the influence index. It is convenient as the simplest measure of dispersion also since it is determined analogously to the influence index of a stochastic parameter (i.e., the regular dispersion σ^2), which allows comparing the parameter x_i directly with the effect of randomness.

We will start with an analysis of one parameter x which is being studied at the values $x_1, x_2, \ldots, x_k$. In order to give the evaluation of response of the other parameter y_i to the prescribed change of x_i, it is necessary to have duplicate observations at each value $x_{i1}, x_{i2}, \ldots, x_{ik}$. The simplest computations are gained in the case when for each x an equal number of observations is performed. The observations corresponding to the value of parameter x_i we will denote by $y_{i1}, y_{i2}, \ldots, y_{in}$.

Through y_i we denote the mean value of observations of the parameter y at the value x_i of the parameter x:

$$y_i = (1/n)\sum_{j=1}^{n} y_{ij}$$

The average of all observations:

$$\bar{y} = 1/(kn)\cdot \sum_{i=1}^{n}\sum_{j=1}^{n}; \quad y_{ij} = (1/k)\sum_{i=1}^{k} y_i$$

General selective dispersion of all the observations

$$S^2 = 1/(kn-1)\sum_{i=1}^{n}\sum_{j=1}^{n}(y_{ij}-\bar{y})^2$$

This dispersion depends on all values of the parameter x and also on factor of randomness at the prescribed value of x_i. The factor of randomness may be evaluated from the random (selective) dispersions characterizing the factor of randomness at all values of x_i,

$$S_i^2 = 1/(n-1)\sum_{i=1}^{n}(y_{ij}-y_i)^2$$

If there are no significant differences between the dispersions S_i^2 then they all may be used for evaluation of general dispersion according to the principle of "running measurements." In this case we get

$$S_0^2 = (1/k)\sum_{i=1}^{k} S_i^2$$

Then the approximation for dispersion of the parameter is

$$\sigma_x^2 \approx S^2 - S_0^2$$

Since the effect of the parameter x is most noticeable on the changes of mean y_i, it is possible to estimate the dispersion σ_x^2 more accurately, namely:

$$1/(k-1)\sum_{i=1}^{k}(y_i-\bar{y})^2 \approx \sigma_x^2 - \sigma^2/2 \approx \sigma_x^2 + S_0^2/n$$

$$\sigma_x^2 \approx 1/(k-1)\sum_{i=1}^{k}(y_i-\bar{y})^2 - S_0^2/n = S_x^2 - S_0^2/n$$

If the influence of the parameter x on y is significant, then dispersion S_x^2 must differ significantly from S_0^2; S_x^2 and S_0^2 may be compared using the one-sided Fisher's criterion (variance ratio) (since $S_x^2 \gg S_0^2$) with $f_1 = k - 1$ and $f_2 = k(n - 1)$ degrees of freedom. If at the selected level of significance p the ratio $S_x^2/S_0^2 > F_{1-p}$, then the influence of parameter x is recognized as significant. If $S^2/S_0^2 \leq F_{1-p}$, then the general dispersion is related only to the factor of randomness.

The described dispersion analysis is used for determination of the set of parameters $x_1, x_2, \ldots, x_n$ at the selected set $y_1, y_2, \ldots, y_m$. If, for example, it turns out that the clearance does not show a considerable effect on lubricant viscosity, then it is obvious that in diagnosis of a bearing the viscosity cannot be used to determine the clearance. This means that some other parameter of the state of lubricant (for example, temperature, contamination, the change of chemical structure) should be chosen which is correlated with the clearance.

To compare influence of the two parameters x_1 and x_2 on the state parameter y, it is necessary to carry out parallel observations for each combination of values x_1 and x_2. Let us denote these observations by y_{ijk}. Here the first subscript indicates the ordinal number of values of the parameter x_1, the second subscript shows the ordinal number of values of the parameter x_2; k is the ordinal number of observation at the state of i-th value of x_1 and of j-th value of the parameter x_2.

Each series of observations has its own dispersion:

$$S_{ij}^2 = k/(n-1) \sum_{k=1}^{n} (y_{ijk} - y_{ij})^2; \quad x_{ij} = 1/k \sum_{i=1}^{k} y_{ijk}$$

Weighted average dispersion across all series is

$$S^2 = 1/(m-k) \sum_{i=1}^{k} \sum_{j=1}^{m} S_{ij}^2$$

The dispersion of the effect of parameters x_1 and x_2 from the equality $S_0^2 = \sigma_{x_1x_2}^2 + (\sigma^2/m)$ can be found from the formula $\sigma_{x_1x_2}^2 \approx S_0^2 - (S^2/n)$. For its estimation we compute

$$Q = \sum_{i=1}^{k} \sum_{j=1}^{m} \sum_{v=1}^{n} y_{ijv}^2; \quad Q_1 = \sum_{i=1}^{k} \sum_{j=1}^{m} y_{ij}^2; \quad Q_2 = 1/m \sum_{i=1}^{k} y_i^2$$

$$x_i = \sum_{j=1}^{m} y_{ij}; \quad Q_3 = 1/k \sum_{j=1}^{m} y_j^2; \quad x_j = \sum_{i=1}^{k} y_{ij}$$

$$Q_4 = 1/(mk) \left(\sum_{i=1}^{k} y_i \right)^2; \quad S_{x_1}^2 = (Q_2 - Q_4)/(k-1)$$

$$S_{x_2}^2 = (Q_3 - Q_4)/(m-1)$$

For the selected level of significance p we will check the correctness of inequality

$$(S^2_{x_1}/S^2_c) > F_{1-p}, \quad S^2_c(Q_1 + Q_4 - Q_2 - Q_3)/[(k-1)(m-1)]$$

Here in F (Fisher's)-distribution $f_1 = k - 1, f_2 = (k - 1)(m - 1)$ degrees of freedom are taken. If the inequality is complied with, then the influence of the parameter x_1 is significant.

In a similar way, the influence x_2 is significant if

$$S^2_{x_2}/S^2_0 > F_{1-p}$$

We determine the interaction of parameters x_1 and x_2 by computing

$$S^2 = (Q_3 - nQ_1)/[mk(n-1)]$$

and comparing nS_0^2 and S^2 according to Fisher's criterion (ratio test). If the difference of these dispersions is significant, the effect of interaction of parameters x_1 and x_2 is also significant.

After selection of basic parameters x_1, structural parameters y_1 and state parameters y_j, the measurement of parameters x_i should be carried out. This experimental material, together with data acquired by extraction from the fundamental parameters, is used for laying down rules for making decisions on the state of a bearing.

Let us formulate rules for making decisions referring to the methods of a maximum likelihood. Here it is necessary from the experimental data to evaluate the probabilities $p(y_j)$ of each of the possible states y_i of a bearing and conditional probability $p_{yj}(x_i)y_j$ at the given structural parameter x_i. We determine them according to the known laws of probability theory and consider them as functions of states of a bearing, i.e., we have the likelihood functions.

The probability of appearance of the structural parameter x_i as a result of experiment we estimate from the Baye formula

$$P_x(y) = 1/P(y) \; P(y) \; P_y(x)$$

In practical applications of these probabilities, in order to make decisions, we compute the values λ_v ($v = 1, 2, \ldots, n$) of the parameter η as logarithms of ratios of á posteriori probabilities

$$\lambda_v = \ln[P_x(y)/P_x(y_1)] = \ln[P(y)L_x/P(\lambda)L_1] = P_x(y)$$

We chose such a structural state of the bearing or, respectively, the value of the parameter x_v, $v = 1, 2, \ldots, k$ for which the value λ_v, $v = 1, 2, \ldots, n$ is maximum and positive. In case all λ_v are negative, we choose the state x_1.

4.2 DETERMINATION OF PROBABILISTIC CHARACTERISTICS

We will consider the probabilistic characteristics in a specific unit. In order to determine axial vibrations of a rotor system hanging on one precision thrust ball bearing there was assembled a setup (Fig. 4.1). The investigation was carried out for a slowly and uniformly rotating outer ring of a bearing ($3 \cdot 10^{-2}$ s^{-1}). The signal from an inductive pickup was amplified by an amplifier DISA 51800 and recorded by a fast response recorder. The obtained plots of axial vibrations were processed by computers.

After the measurement of vibrations the bearings were disassembled and deviations from the theoretical circumference of races of inner and outer rings due to inaccurate manufacturing and treatment were recorded on a Talyrond. The worn part of the race was found by the charging method. Figure 4.2 shows the Talyrond profilograms.

Figure 4.3 shows the plots of vibrations of the rotor system in an axial direction. Development of oscillations in rolling contact bearings is due to the following reasons: errors of mounting; contamination; errors in manufacturing of the shaft and the bearing itself. According to Perrett, vibrations under steady-state loading as a result of elastic deformation of races may even arise when there are no faults in manufacture and mounting. Meldau also explains his theory of development of vibrations by deformation, or by so-called Hertzian compression.

In order to get a better picture, the investigation was carried out using computers. The study was intended to obtain spectral densities and correlation functions of axial oscillations of the rotor system.

The functions of axial vibrations of a rotor system obtained from Hemming's formula are shown in Fig. 4.4.

Experimental studies of axial oscillations of a rotor system allow making an

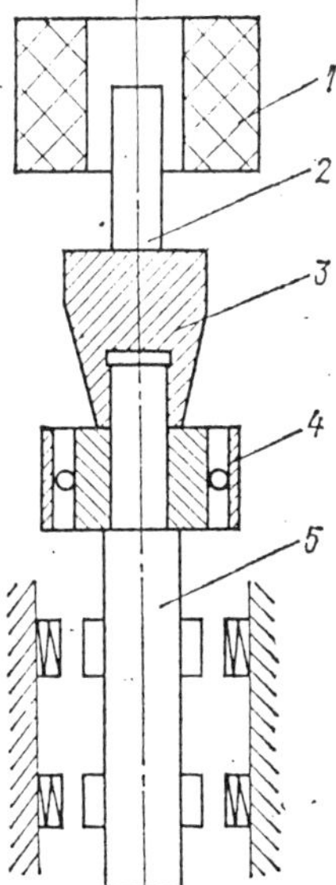

Figure 4.1 Schematic of setup for determining vibration characteristics of a bearing: *1*) stator of variance reluctance pickup; *2*) rotor of variance reluctance pickup; *3*) collet clamp; *4*) test bearing; *5*) axle.

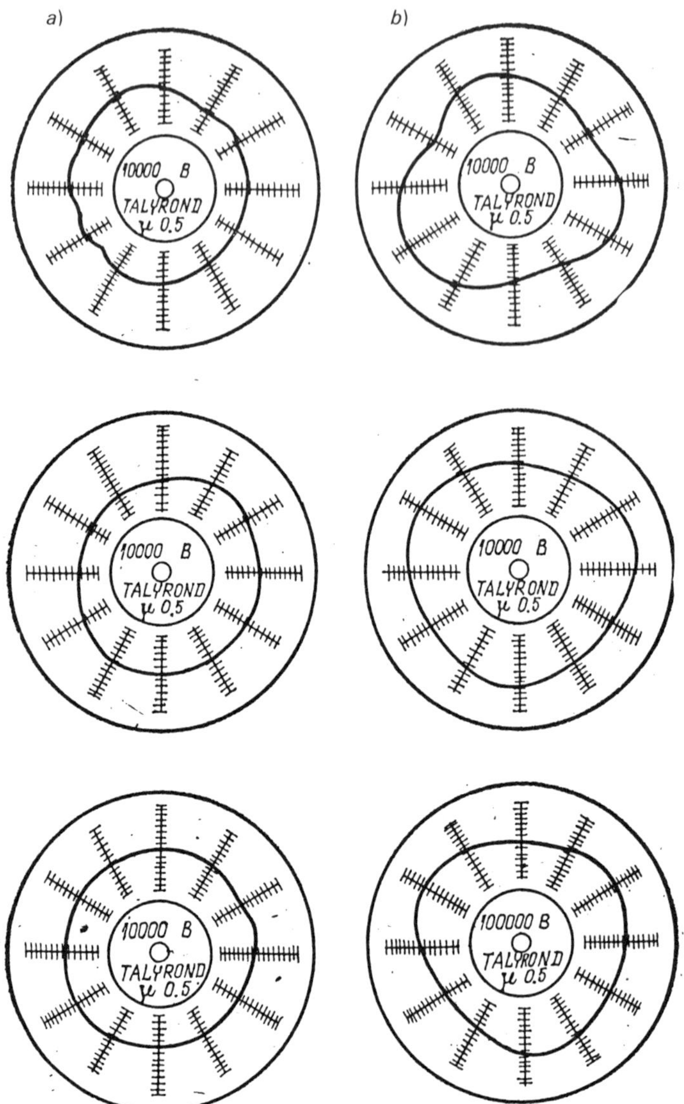

Figure 4.2 Examples of roundness diagrams describing deviations from correct geometrical shape of rings (of bearings *1*, *2*, *3*): *a*) inner ring; *b*) outer ring.

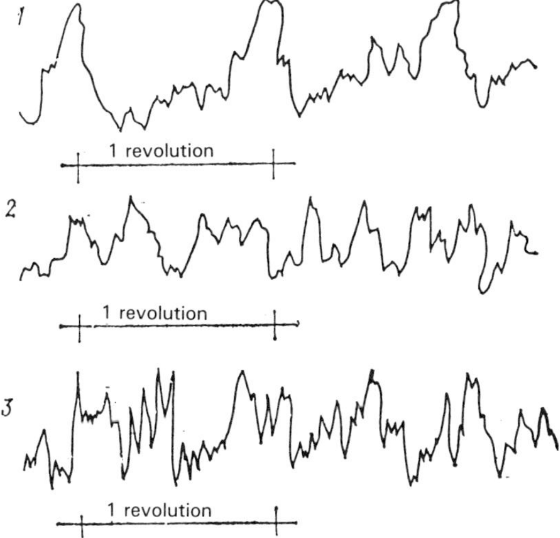

Figure 4.3 Vibration of rotor system in axial direction: *1, 2, 3* are numbers of bearings subjected to tests.

analytical computation of spectral density of these oscillations by assuming geometric errors of races as correlation functions.

Vibrations of the inner ring of a bearing (in this case also of the entire rotor system) are produced mainly by two sources: 1) by cyclic changes of bearing compliance under the action of loading, which arise even in the case of a bearing with perfect geometry, and 2) vibrations due to geometric errors of surfaces in contact with rolling bodies during rolling.

Other sources of vibration (contact between the balls, mounting misalignment, nonuniformity of materials, etc.) increase relative ring vibrations insignificantly. In angular ball bearings, races are in contact with the balls along the angle α. The generalized model of a mechanical system we will represent as shown in Fig. 4.5. The bearing system may be described by equations:

$$\left.\begin{aligned} m_1\ddot{x}_1+k_1[x_1+\varphi_1(t)]+c_1\dot{x}_1+k_2[x_1-x_2+\varphi_2(t)]+c_2(\dot{x}_1-\dot{x}_2)&=0 \\ m_2\ddot{x}_2+k_2[x_2-x_1+\varphi_2(t)]+c_2(\dot{x}_2-\dot{x}_1)&=0 \end{aligned}\right\}$$

Here m_1 and m_2 are the masses of ball and inner ring of a bearing, respectively; c_1 and c_2 are damping coefficients of outer ring (mass)—ball and ball—and inner ring subsystems, respectively; k_1 and k_2 are stiffness in subsystem's outer ring—ball and ball—and inner ring, respectively; $\varphi_1(t)$, $\varphi_2(t)$ are the profiles of geometric errors of outer ring and inner ring, respectively.

After the simple transformations, the system becomes

$$\ddot{x}_1+[(c_1+c_2)/m_1]\,\dot{x}_1-(c_1/m_1)\,x_1+(c_2/m_1)\,\dot{x}_2-(k_2/m_1)\,x_2=(k_1/m_1)\times\varphi'(t)-(k_2/m_1)\,\varphi_2(t)$$

$$\ddot{x}_2+(c_2/m_2)\,\dot{x}_2+(k_2/m_2)\,x_2-(c_2/m_2)\,\dot{x}_1-(k_2/m_2)\,x_2=-(k_2/m_2)\,\varphi_2(t)$$

Let us assume that $\varphi_1(t)$ and $\varphi_2(t)$ are normal, stationary, random functions. This system will solve with zero initial conditions. Of interest are the probabilistic characteristics of random process x_2. We will assume that mathematical expectations for $\varphi_1(t)$ and $\varphi_2(t)$ are zeros, and correlation functions are equal to:

$$k\varphi_1(\tau) = \sigma^2\varphi_1 e^{-\alpha(\tau)}; \quad k\varphi_2(\tau) = \sigma^2\varphi_2 e^{-\alpha(\tau)} \cos\beta\tau$$

where $\sigma^2\varphi_1$ and $\sigma^2\varphi_2$ are the dispersions φ_1 and φ_2; $\alpha > 0$ has a unit of measurement which is inverse to the unit of time measurement and may serve as a characteristic of rate of correlation decrease between the ordinates of random function at the increase of difference in arguments of these ordinates $\tau \cos\beta\tau$. Then $k\varphi_2(\tau)$ will assume the form of decaying harmonic oscillation.

We will assume that $\varphi_1(t)$ and $\varphi_2(t)$ are not associated with each other, i.e., the cross correlation function $K\varphi_1\varphi_2(\tau) = 0$. In this case, if we denote $\varphi_1(t) =$

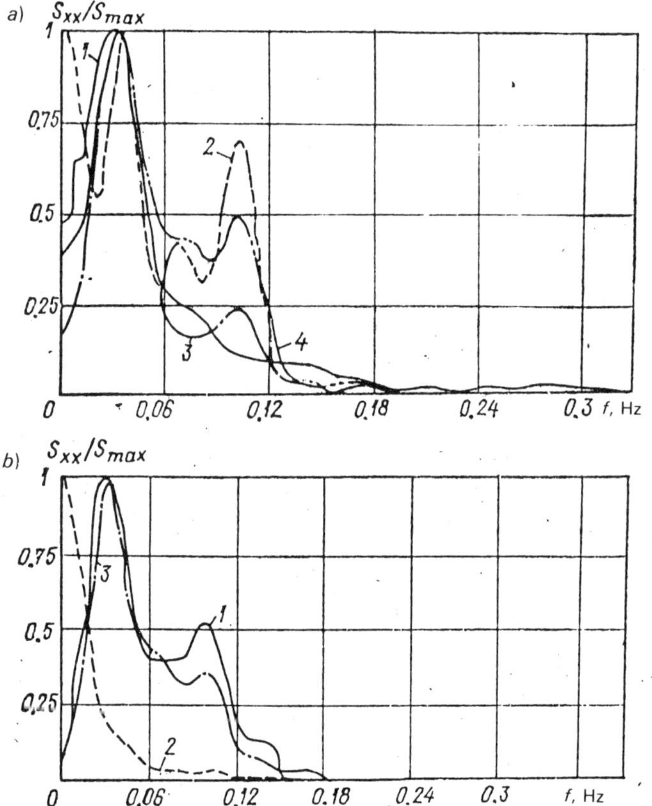

Figure 4.4 Spectral densities of axial vibration of rotor system of bearings: *a*) dry; *b*) lubricated; *1, 2, 3, 4*) curves corresponding to the bearing numbers.

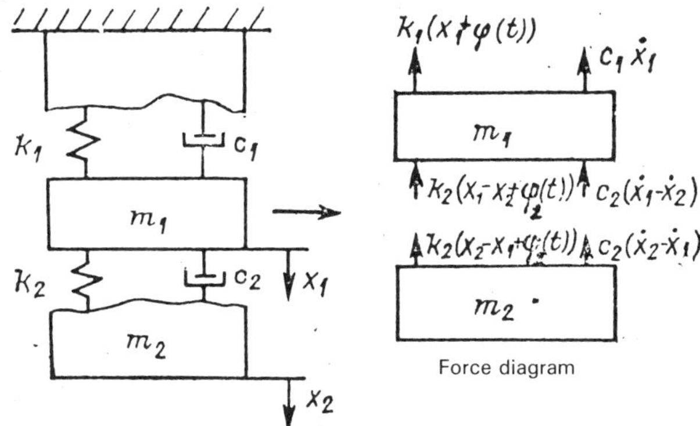

Figure 4.5 Generalized model of a mechanical system.

$(K_1/m_1)\varphi_1(t) - (K_2/m_1)\varphi_2(t)$ and $\varphi_2(t) = -(K_2/m_2)\varphi_2(t)$ and take into account that $-(K/m_1)$; $-(K_2/m_1)$ and $-(K_2/m_2)$ are constants, we will get:

$$K_{f_1}(\tau) = K_1/m_1^2\sigma\varphi_1 e^{-\alpha(\tau)} + K_2^2/m_1^2\sigma_1^2\varphi_2 e^{-\alpha(\tau)}\cos\beta\tau$$

$$K_{f_2}(\tau) = K_2^2/m_1^2\sigma^2\varphi_2 e^{-\alpha(\tau)}\cos\beta\tau$$

Still, it is necessary to determine the cross correlation function between $f_1(t)$ and $f_2(t)$.

As it is known, if two random functions have the form:

$$y_1(t) = a_1(t)x_1(t) + a_2(t)x_2(t); \quad y_2(t) = b_1(t)x_1(t) + b_2(t)x_2(t)$$

where a_1, b_1, a_2, b_2 are nonrandom functions of time, then

$$K_{y_1y_2}(t_1t_2) = a_1^*(t_1)b_1(t_2)K_{x_1}(t_1t_2) + a_2^*(t_1)b_2(t_2)K_{x_2}(t_1t_2)$$
$$+ a_1^*(t_1)b_2(t_2)K_{x_1x_2}(t_1t_2) + a_2^*(t_1)b_1(t_2)K_{x_2x_1}(t_1t_2)$$

Here a_1^*, a_2^* are the complex functions conjugate with a_1 and a_2.

In our case:

$$f_1(t) = -(K_1/m_1)\varphi_1(t) - (K_2/m_1)\varphi_2(t)$$
$$f_2(t) = -(K_2/m_2)\varphi_2(t)$$

Since the coefficients in the right-hand parts are constant and stationary, then, using the general formula we get

$$K_{f_1f_2}(\tau) = -(K_2/m_1)\left[-(K_2/m_2)\right]K\varphi_2(\tau) = K_2^2/(m_1m_2)\,\sigma^2\varphi_2 e^{-\alpha(\tau)}\cos\beta\tau$$

$$K_{f_1f_2}(\tau) = K_{f_1f_2}(-\tau); \qquad \text{in our case} \qquad K_{f_2f_1} = K_2^2/(m_1m_2)$$
$$\times\,\sigma^2\varphi_2 e^{-\alpha(\tau)}\cos\beta\tau, \quad \text{such that} \quad K_{f_1f_2}(\tau) = K_{f_2f_1}(\tau)$$

The set of differential equations we will rewrite in a canonical form with the following notations:

$$(c_1 + c_2)/m_1 = a_1; \quad (k_1 + k_2)/m_1 = a_2; \quad c_2/m_1 = a_3; \quad k_2/m_1 = a_4.$$
$$c_2/m_2 = b_1; \quad k_2/m_2 = b_2; \quad dx_1/dt = y_1(t); \quad x_1 = y_2(t)$$
$$dx^2/dt = y_3(t); \quad x_2 = y_4(t)$$
$$\dot{y}_1 + a_1 y_1 + a_2 y_2 - a_3 y_3 - a_4 y_4 = f_1(t)$$
$$\dot{y}_2 - y_1 = 0$$
$$\dot{y}_3 + b_1 y_3 + b_2 y_4 - b_1 y_1 - b_2 y_2 = f_2(t)$$
$$\dot{y}_4 - y_3 = 0$$

Since the system is stable and the time t is large enough, the transient process may be considered as completed. Then the spectral densities for random functions $y_j(t)$ are derived from the general formula:

$$S_{yj}(\omega) = \sum_{p=1}^{n} [A_{je}(\omega)]^2 S_{f_2}(\omega) + \sum_{\substack{e=1 \\ e \neq k}}^{n} \sum_{k=1}^{n} A^*_{je}(\omega) A_{jk}(\omega) S_{f_e f_k}(\omega)$$

In this case for y_4 we get

$$S_{y_4}(\omega) = S_{f_2}(\omega) = [A_{41}(\omega)]^2 S_{f_1}(\omega) + [A_{42}(\omega)]^2 S_{f_2}(\omega) + A^*_{41}(\omega) \times A_{42}(\omega) S_{f_1 f_2}(\omega) + A^*_{42}(\omega) A_{41}(\omega) S_{f_1 f_2}(\omega)$$

where $A_{je} = \Delta ej/\Delta$; Δ is the system determinant; Δej is the algebraic adjunct of this determinant corresponding to the element at the intersection of the e-th row and the j-th column;

$$\Delta = \begin{vmatrix} a_1 + i\omega & a_2 & -a_3 & -a_4 \\ -1 & i\omega & 0 & 0 \\ -b_1 & -b_2 & b_1 + i\omega & b_2 \\ 0 & 0 & -1 & i\omega \end{vmatrix}$$

$$\Delta = (\omega^2 - a_1\omega_i - a_2)(\omega^2 - b_1\omega_i - b_2) - (a_4 + a_3\omega_i)(b_2 + b_1\omega_i)$$
$$\Delta = \omega^4 - a_1\omega_i^3 - a_2\omega^2 - b_1\omega_i^3 - a_1 b_1\omega^2 + b_1 a_2\omega_i - b_2\omega^2 + a_1 b_2\omega_i$$
$$+ a_2 b_2 - b_2 a_4 - b_2 a_3\omega_i - b_1 a_4\omega_i + b_1 a_3\omega^2 = \omega^4 + (b_1 a_3 - a_2 - a_1 b_1$$
$$+ b_2)\,\omega^2 + a_2 b_2 + b_2 a_4 - (a_1 + b_1)\,\omega_i^3 + (b_1 a_2 + a_1 b_2 + b_1 a_3 - b_1 a_4)\,\omega_i$$

If we accept the following notations for brevity:

$$b_1 a_3 - a_1 b_1 - a_2 - b_2 = \alpha_1; \quad a_2 b_2 - b_2 a_4 = \beta_1; \quad a_1 + b_1 = \gamma$$
$$b_1 a_2 + a_1 b_2 - b_2 a_3 - b_1 a_4 = \delta_1,$$

then

$$\Delta = \omega^4 + \alpha_1\omega^2 + \beta_1 + \gamma\omega_i^3 + \delta_{\omega_i}$$

$$\Delta_{14} = \begin{vmatrix} -1 & i\omega & 0 \\ -b_1 & -b_2 & b_1 + i\omega \\ 0 & 0 & -1 \end{vmatrix} = \begin{vmatrix} -1 & i\omega \\ -b_1 & -b_2 \end{vmatrix} = b_2 + b_1\omega_i$$

$$\Delta_{24} = \begin{vmatrix} a_1 + i\omega & a_2 & -a_3 \\ -b & -b_2 & b_1 + i\omega \\ 0 & 0 & -1 \end{vmatrix} = -\begin{vmatrix} a_1 + i\omega & a_2 \\ -b_1 & -b_2 \end{vmatrix}$$

$$= a_1b_2 - b_1a_2 + b_2\omega_i$$

$$A_{41} = \Delta_{14}/\Delta = (b_2 + b_1\omega_i)/(\omega^4 + \alpha_1\omega^2 + \beta_1 - \gamma\omega_i^3 + \delta_{\omega_i})$$

$$A_{41}^* = (a_1b_2 - b_2a_2 - b_2\omega_i)/(\omega^4 + \alpha_1\omega^2 + \beta_1 - \gamma\omega_i^3 + \delta_{\omega_i})$$

$$A_{42} = \Delta_{24}/\Delta = (a_1b_2 - b_1a_2 + b_2\omega_i)/(\omega^4 + \alpha_1\omega^2 + \beta_1 - \gamma\omega_i^3 - \delta_{\omega_i})$$

$$A_{42}^* = (b_2 - b\omega_i)/(\omega^4 + \alpha_1\omega^2 + \beta_1 - \gamma\omega_i^3 - \delta_{\omega_i})$$

It is still necessary to obtain the expression for $S_{f_1}(\omega)$; $S_{f_2}(\omega)$ and $S_{f_1f_2}(\omega)$; which are as follows

$$S_{f_1}(\omega) = K_1^2\sigma\varphi_1/(2\pi m_1^2)\,[1/(\alpha + \omega_i) + 1/(\alpha - \omega_i)]$$
$$+ K_2^2\sigma^2\varphi_2/(2\pi m_1^2)\,\{[\alpha/(\omega - \beta)^2 + \alpha^2] + [\alpha/(\omega + \beta)^2 + \alpha^2]\}$$

$$S_{f_2}(\omega) = K_2^2\sigma\varphi^2/(m_2^2 2\pi)\int_{-\infty}^{\infty} e^{-i\omega\tau}\, e^{-\alpha\tau}\cos\beta\tau\; d\tau = K_2^2\sigma\varphi_2\alpha/(m_2^2\pi)$$
$$\times\{(\omega^2 + \alpha^2 + \beta^2)/[(\omega^2 - \alpha^2 - \beta^2)^2 + 4\alpha^2\omega^2]\}$$

$$S_{f_1f_2}(\omega) = K_2^2\sigma^2\varphi_2\alpha/(m_1m_2\pi)\,\{(\omega^2 + \alpha^2 + \beta^2)/[(\omega^2 - \alpha^2 - \beta^2) + 4\alpha^2\omega^2]\}$$

Since $S_{f_2f_1}(\omega) = S_{f_1f_2}(\omega)$, we get

$$S_{f_2f_1}(\omega) = K_2^2\sigma^2\varphi_2\alpha/(m_1m_2\pi)\,[(\omega^2 + \alpha^2 + \beta^2)/(\omega^2 - \alpha^2 - \beta^2) + 4\alpha^2\omega^2]$$

It is possible now to obtain the final expression for the spectral density

$$S_{x_2}(\omega) = \alpha/(\pi m_1^2) - \{(b_2^2 + b_1^2\omega^2)/[(\omega^4 + \alpha_1\omega + \beta_1)^2 + (\gamma\omega^2 - \delta)^2\omega^2]\}$$
$$\times\{K_1^2\sigma^2\varphi_1[(\omega^2 - \alpha^2 - \beta^2)^2 + 4\alpha^2\omega^2] + [K_2^2\sigma^2\varphi_2(\omega^2 + \alpha^2 + \beta^2)(\omega^2$$
$$- \alpha^2)]/(\omega^2 + \alpha^2)[(\omega^2 - \alpha^2 - \beta^2)^2 + 4\alpha^2\beta^2]\} + \{K_2^2\sigma^2\varphi_2\alpha/(\pi m_1^2)[(a_1b_2$$
$$- a_2b_1)^2 + b_2^2\omega^2](\omega^2 + \alpha^2 + \beta^2)\}/\{[(\omega^4 + \alpha_1\omega^2 + \beta_1)^2 + (\gamma\omega^2 - \delta)$$
$$\times\omega^2][\omega^2 - \alpha^2 - \beta^2) + 4\alpha^2\beta^2]\} + [(2K_2\sigma^2\varphi_2\alpha)/(m_1m_2\pi)]\{[(a_1b_2^2$$
$$- a_2b_1b_2) + b_1b_2\omega^2]/[(\omega^4 + \alpha_1\omega^2 + \beta_1)^2 + (\gamma\omega^2 - \delta)^2\omega^2][(\omega^2 - \alpha^2$$
$$- \beta^2)^2 + 4\alpha^2\beta^2]\}$$

The correlation function for x_2 may be obtained from the correlation

$$K_{x_2}(\tau) = \int_{-\infty}^{\infty} e^{i\omega\tau} S_{x_2}(\omega)\, d\omega$$

after the substitution of the obtained expression for $S_{x_2}(\omega)$. In the right-hand side of the latter equality, three intervals will be obtained in the form

$$\int_{-\infty}^{\infty} \{[P_m(i\omega)]^2/[Q_n(i\omega)]^2\}\, e^{i\omega\tau}\, d\omega$$

Since under our assumptions $\varphi_1(t)$ and $\varphi_2(t)$ are normal stationary random functions, then x_2 is also a normal random stationary function. If we consider that the spectral density for $x_2(t)$ has a sharp maximum, then the spectrum may be approximately regarded as a narrow band one. At such conditions $x_2(t) = A(t) \cos[\omega_1 t + \theta(t)]$, where $A(t)$ is the amplitude of the envelope; $\theta(t)$ is the deviation from the phase of harmonic oscillatory motion; $A(t)$ and $\theta(t)$ are the random functions.

The correlation function for $A(t)$ has the form

$$K_Q(\tau) = \sigma_{x_2}^2 [2E(1-p^2) - p^2 K(1-p^2) - \pi/2]$$

where

$$K(1-p^2) = \int_0^{\pi/2} d\varphi / \sqrt{1-(1-p^2)\sin^2\varphi}; \qquad E(1-p^2)$$

$$= \int_0^{\pi/2} \sqrt{1-(1-p^2)\sin^2\varphi}\, d\varphi; \qquad p^2 = 1 - K^2(\tau) - r^2(\tau); \qquad K(\tau)$$

$$= 1/(\sigma_{x_2}^2) K_{x_2}(\tau); \quad \tau = 2\int_{-\infty}^{\infty} S_{x_2}(\tau) \sin\omega\tau\, d\omega$$

$S_{x_2}(\omega)$ is the normalized spectral density related to $\sigma_{x_2}(\omega)$ as $S_{x_2}(\omega) = 1/(\sigma_{x_2}^2) S_{x_2}(\omega) > (\tau)$, and may be obtained by the numeric method:

$$W = 1/(\sigma^2 x_2) \int_{-\infty}^{\infty} (\omega)\, S_{x_2}(\omega)\, d\omega$$

This derivation enables an analysis of spectral/correlation characteristics of bearing rings' vibrations when the correlation functions of races' geometric shape are known; the latter may be different for each specific case. More complete characteristics of the system performance may be obtained if it is described by a nonlinear equation (by a set of nonlinear equations). Statistic characteristics of solutions of nonlinear equations with the aid of analytic methods are determined accurately only in simple cases, hence it is preferable to resort to numeric methods. The algorithms based on the Monte Carlo method have the simplest computational scheme. They were used in determining the statistic characteristics of rotor vibrations in radial direction.

It is assumed that a vertical rigid rotor is ideally balanced and rotates with constant speed in a rolling contact bearing, and the oscillations of the rotor are possible only in radial direction. Dissipation of energy in the bearing is assumed to be proportional to the speed of oscillating motion of the mass. The bearing stiffness is considered as a function of displacement x of a mobile ring from its position of balance and of the turning angle φ of the cage with the rolling bodies: $-k = k(x\varphi)$. For small values, x can be approximated by expression $k = a - bx^2$, and coefficients a and b are selected for the ball bearings of this standard size according to [5].

The equation of correlation between vibrations x of the rigid balanced shaft in a ball bearing and manufacturing geometric errors in the shape of bearing elements is

$$m\ddot{x} + \beta\dot{x} + ax - bx^3 = \varphi(t) \tag{4.1}$$

where m is the mass of the rotor; $\varphi(t)$ is the random action generated by the microprofiles of bearing elements.

Analysis of the statistic characteristics of actual signals (microprofile functions) has shown that geometric errors have normal distribution (5% criterion χ^2—the significance level), and correlation functions are well approximated by the sum of damped cosines. These signals were simulated on a computer, i.e., there were generated random sequences possessing characteristics analogous to actual signals. The pseudorandom uniformly distributed numbers were generated, which were further transformed by a nonlinear asymptotically normal transformation. The obtained numerical correlation sequence ξ_i was transformed into a sum of signals $\varphi_k(i\ \Delta t)$

$$\varphi(i\Delta t) = \sum_{k=1}^{n} \varphi_k(i\Delta t) \tag{4.2}$$

which have a correlation function in the form

$$K_{\varphi_k}(\tau) = \sigma_k^2 e^{-\alpha k(\tau)} \cos \beta_k(\tau)$$

Here Δt is the step of digitization in time for numerical determination of the signal $\varphi(t)$.

Using the special case of Runge-Kutt formula, which has an order of error at one step which equals the fifth power of a step of integration (the step of digitization Δt), the integration of differential equation (4.1) was carried out at zero initial conditions and with specific numerical values of the random function $\varphi(t)$, which were determined by the above-described method. In order to cut the computation time and to obtain the solution of the system having a relative error not exceeding the prescribed level, the selected step of integration is variable. The obtained sequence of values of the solution of the equation is reduced to the sequence determined in the points with uniform spacing. This requires the use of linear interpolation.

A multiple (10^3–10^4 times) repetition of reproducing the signal $\varphi(t)$ and solving the equation (4.2) for different samplings of random functions $\varphi(t)$ provides the statistical material in the form of $x_{2k}(t)$, where $k = 1, 2, \ldots, N$ is the number of simulations.

The values of the required statistical characteristics of the equation (4.1) can be obtained from the formula

$$M[\Phi] = (1/N) \sum_{k=1}^{N} \Phi_k$$

where Φ_k is the required characteristic for the simulation k. In our case it is a mathematical expectation, correlation function, spectral density, and function of distribution of signal x_2, describing vibrations of the shaft in a ball bearing.

4.3 DETERMINATION OF STIFFNESS COEFFICIENTS

The precise determination of the stiffness of various components, units, and structures is the kind of job which consumes much time. There exists a great number of ways for stiffness measurement. Here the characteristics of stiffness are used as the diagnostics parameters. The radial-thrust ball bearing assemblies with axial preload are considered here as an example of stiffness determination. This method may be used also for other components, units, and structures. In this case, the stiffness is considered linear. The determined stiffness coefficients correspond to the actual stiffness of the system being measured. In systems having two degrees of freedom there are two coefficients of stiffness. Let us analyze the system consisting of a rigid shaft with two radial-thrust ball bearings (Fig. 4.6).

The equations of motion of such a system are:

$$\left.\begin{aligned} m\ddot{x} + (k_1 + k_2)\,x - (x_1 L_1 - k_2 L_2)\,\theta = \theta_1 \\ -(k_1 L_1 - k_2 L_2)\,x + J\theta + (k_1 L_1^2 + k_2 L_2^2)\,\theta = \theta_1 \end{aligned}\right\} \tag{4.3}$$

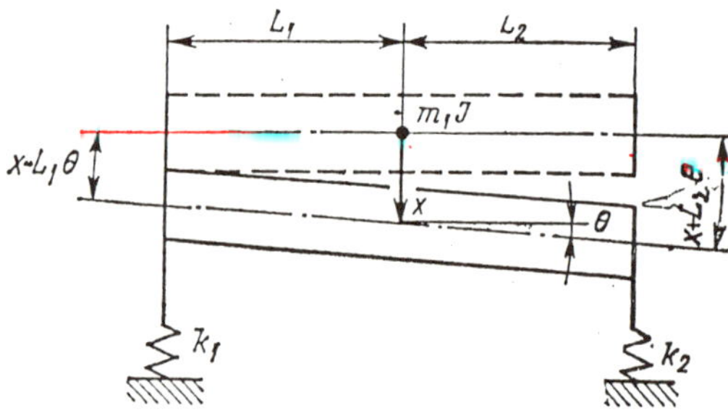

Figure 4.6 Schematic for computing stiffness of the system by the search method.

where k_1 and k_2 are radial stiffness coefficients of the first and the second ball bearings; m is the mass of the moving part; J is inertia moment; L_1 and L_2 are distances from the center of gravity of the axle to the vertical planes passing through the ball centers of bearings; x and θ are the motion coordinates of the system being measured.

The equations (4.3) are linear differential equations with constant coefficients. Their solution is

$$x = A_1 \sin(\omega t + f); \quad \theta = A_2 \sin(\omega t + f) \tag{4.4}$$

We will substitute Eq. (4.4) into (4.3), and after algebraic transformations obtain the natural frequency equation as

$$\omega^4 - \{[(k_1 - k_2)/m] + [(k_1 L_1^2 + k_2 L_2^2)/J]\}\,\omega^2 + \{k_1 k_2 (L_1 + L_2)^2/Jm\} = 0 \tag{4.5}$$

This is a square equation relative to ω^2, whose solution is

$$\omega^2 = 1/2\{[(k_1 + k_2)/m] + [(k_1 L_1^2 + k_2 L_2^2)/J] \pm \sqrt{[(k_1 + k_2)/m + (k_1 L_1^2 + k_2 L_2^2)/J]^2 - 4[k_1 k_2 (L_1 + L_2)^2/(Jm)]}\}$$

Natural frequencies ω_1 and ω_2 of the systems being measured are

$$\left.\begin{aligned} \omega_1 &= \sqrt{1/2\{(k_1 + k_2)/m + (k_1 L_1^2 + k_2 L_2^2)/J + \sqrt{[(k_1+k_2)/m+(k_1 L_1^2+k_2 L_2^2)/J]^2-4[k_1 k_2(L_1+L_2)^2/(Jm)]}\}} \\ \omega_2 &= \sqrt{1/2\{(k_1 + k_2)/m + (k_1 L_1^2 + k_2 L_2^2) J - \sqrt{[(k_1+k_2)/m+(k_1 L_1^2 + k_2 L_2^2)/J]^2-4[k_1 k_2 (L_1+L_2)^2/(Jm)]}\}} \end{aligned}\right\} \tag{4.6}$$

The systems with two degrees of freedom have two modes of oscillations and only under limited conditions does each system perform oscillations in accordance with one of these modes. In case of resonance of forced vibrations, the modes of forced vibrations are analogous to the corresponding principal modes. In this case the resonance frequencies are equal to the natural frequencies of the system.

The stiffness coefficients of the system (in this case with radial-thrust ball bearings) will be determined by means of a search by computer. It may be assumed that the stiffness coefficients k_1 and k_2 are between the limit boundaries: $k_1' < k_1 < k_1''$; $k_2' < k_2 < k_2''$.

It is necessary to find k_1 and k_2, which are the stiffness coefficients of the system being measured. The closeness function is compiled for assessment of the search accuracy

$$F = \sum_{i=1}^{n} (\omega_i - \omega_i^*)^2$$

Here ω_i are search natural frequencies of the system determined from the equations (4.6) while the stiffness coefficient values are being changed; ω_i^* are experimental natural frequencies of the system which were determined by the resonance method; $i = 1, 2, \ldots, n$.

In the given case, the closeness function is

$$F = (\omega_1 - \omega_i^*)^2 + (\omega_2 - \omega_2^*)^2$$

When the closeness function is equal to zero, the stiffness coefficients k_1 and k_2 are selected correctly and serve as the stiffness coefficients of the test system.

The initial values k_1' and k_1'' as well as k_2' and k_2'' could be found from the experimental tables knowing the type and the class of the bearing accuracy or by formulas. In our case for radial-thrust ball bearings, the formula was used

$$k = F \cos \beta/(2\omega \sin \beta)[3/2 \cos \beta + \omega/s \sin \beta]$$

where F is the reaction force acting on the inner ring of the bearing; ω is the deformation of balls and races; s is the distance between the centers of the curvature of grooves; β is the contact angle in a bearing.

The calculations carried out and the experimental investigations have shown that the method for determining stiffness coefficients provides accurate results.

4.4 STATISTICAL PROCESSING TECHNIQUE OF EXPERIMENTAL RESULTS

Before an analysis of statistical characteristics of parameter realizations of bearing units, it is necessary to establish the principal characteristics of the random process. These are degree of randomness, stationarity, and distribution density. The techniques for obtaining statistical characteristics are detailed in [5, 9, 22].

The verification of a randomness of the process is reduced to verification of the presence of harmonic components, due to the presence in the process of periodic or nearly periodic components. Harmonic components having large power levels usually are noticed at once. However, the presence of harmonic components with small energy may be less obvious. The most effective in detecting periodic components are the same methods which are applied for analysis of other characteristics of random processes. Therefore, the practical criterion of randomness is based on methods of analysis which are used when the processes are assumed to be random. For example, the presence of harmonic components in random, harmonic, and nonharmonic processes is possible to detect while viewing the graphs of spectral density, probability density, or autocorrelation function obtained for the stationary process.

The most effective method for extracting harmonic vibrations in a random process is plotting its autocorrelation function. In a purely random process the correlation function always approaches zero with an increase in shift (it is assumed that the process is a centered one). On the other hand, the correlation function of harmonic process or sum of harmonics continues to oscillate independent of the magnitude of the shift. To prove stationarity of a random process, it is necessary, theoretically, to verify whether their statistical characteristics are absent or not with a large shift in time. Obviously, in practice such verification is impossible because the number of possible statistical characteristics is infinite and for a complete description of the random process it would be necessary to compute all these characteristics. However, it is possible to suggest a number of practical criteria for a stationary random process.

The first important assumption is that if the processes are not stationary, their statistical characteristics computed on the basis of one realization by averaging along short intervals of time will be different from interval to interval. If the random processes are stationary, then their statistical characteristics computed in short intervals of time would not change appreciably from interval to interval. With this assumption, verification as to whether the random process is stationary consists of the analysis of the behavior of individual realizations and not of the whole ensemble. In other words, with the above assumption, proof that individual realizations are stationary in themselves means that the random process to which these realizations belong is also stationary.

According to the second assumption, for a majority of applications it is sufficient to verify that the process is weakly stationary, i.e., to confirm that the mean values and autocorrelation function do not depend on time. If this assumption is accepted then the verification of the stationarity of a process is accepted to the analysis only of mean values and autocorrelation functions of the processes. This assumption can be accepted for two reasons. First, in order to use such effective methods of analyzing stationary processes as correlation-spectral analysis it is only necessary that the condition of weak stationarity is complied with. Second, random processes observed in actual physical phenomena are usually stationary in a strict sense.

The third assumption states that the realization length is large when compared to the random oscillations contained in this realization. This assumption is necessary so that as a result of averaging at short intervals it is possible to obtain values that correctly reflect the mean characteristics of the process not subjected to the effect of random vibrations contained in this realization, i.e., the length of realization must be so large that it would be possible to tell lengthy periodic components from fluctuations.

According to the fourth assumption, if the mean value of the square (or dispersion) of the process is stationary, the correlation function is also stationary. This assumption is not so very well founded as the first three. However, it is usually assumed because of the small probability of the correlation function of the nonstationary process depending on time for any shift τ except for $\tau = 0$. Since the mean value of the square equals the magnitude of autocorrelation function at $\tau = 0$, the analysis should be confined to the mean value of the square of the process (or dispersion) but not of all values of autocorrelation function.

On the basis of these assumptions it is possible to carry out verification of stationarity of a random process using the examination of one realization $x(t)$ as described below:

1. The realization is divided into N equal intervals in such a way that its segments within each interval can be considered independent. In the case of the process with a relatively wide spectrum, these intervals may be adjacent. But if the process has a narrow spectrum or is formed by the components with low frequency as compared to the length of an individual interval, then it is necessary to leave spaces between the neighboring intervals so that the segments of the process within each interval are totally independent on the neighboring segments.

2. The mean values and mean square values are computed within each interval. From these selected values the sequences of the following form are set up:

$$\bar{x}_1,\ \bar{x}_2,\ \bar{x}_3,\ \ldots,\ \bar{x}_N$$
$$\bar{x}_1^2,\ \bar{x}_2^2,\ \bar{x}_3^2,\ \ldots,\ \bar{x}_N^2$$

3. The sequence of mean values and mean square values is verified for the presence of trends or deviations that exceed the expected values of deviations due to the variation between samples. The final verification of sample values for the presence of a nonstationary trend may be performed in various ways. If the sample distribution of values is known, it is possible to use various parametric criteria. However, in order to find the sample distribution of mean values and dispersions it is necessary to possess a detailed knowledge of frequency composition of the process. Usually, in determining whether the process is stationary, such information is not available. For this purpose it is worth using non-parametric methods which do not require a knowledge of the sample distributions.

Let us assume that the sequences of mean values of processes and mean square values represent the sample values of a random parameter which has a true mean value of the square. If this hypothesis is correct, then the changes of sample values in sequences are random and do not contain a trend. Hence, the number of series in a sequence, where the series are determined with respect to any given value, will be equal to their expected number in a sequence of independent random values being observed. In addition to it, the number of inversions in a sequence will be equal to the expected number of inversions in a sequence of independent random observed values of the same variable. If the number of series or inversions differs considerably from the expected one, then the hypothesis on random process being stationary is to be rejected, if otherwise—the hypothesis is to be accepted.

The obtained correlation functions of the random process show that these functions are decaying. This confirms our assumption that the test random process is also ergodic.

The most obvious way of checking whether the realizations of a stationary random process obey the normal or some other theoretical law is the computation of density distribution of the process values and its comparison with the respective theoretical distribution. If the length of realization is sufficiently large and the errors of the measurement are small as compared to the deviations of the function from the curve having the selected theoretical distribution, then the discrepancy between the selected and theoretical distribution will be obvious. If the sample distribution of density estimate is known, it is possible to use parametric criteria even in the case when statistical errors are large. However (as in the case of stationarity verification) in finding the sample distribution of density estimate it is necessary to know the frequency structure of the process. In practice such information is rarely available. Thus, non-parametric criteria are preferable. One of the most convenient non-parametric criteria for measuring normality of distribution is the agreement criterion χ^2. This criterion should be applied to the discrete sampled values characterizing the random process under test. If the analysis is carried out by numerical methods, no difficulties arise since the process is already represented by the sequence of discrete sample values.

For obtaining the needed statistical characteristics (of correlation function and spectral density) using computers, it is necessary to solve the problems of quantization and digitization, substitution of integrals by sums, and establishment of relation between the duration of realization and frequency resolution and respective parameters of discrete realizations. The statistical errors connected with statistical computations must be determined by these parameters.

In order to obtain accurate information on high frequency components, the amount of sampling moments of time must be rather large. On the other hand, the observations of the points located very close to each other will lead to obtaining correlated data whose number would be redundantly large. This causes increases both in volume and in cost of computations. In order to cut the number of readings, the speed of digitization is reduced to an allowable minimum, thus making it possible to avoid the error of masking. If the time interval Δt between the consecutive readings equals h, then the speed of digitization is $1/h$ readings per second. Besides, useful information may be obtained only for the range of frequencies up to $\frac{1}{2}h$ Hz, because the components with frequencies higher than $\frac{1}{2}h$ Hz will be in the range of low frequencies 0–$\frac{1}{2}h$ Hz being indistinguishable from the components with the same low frequencies. The limit frequency $f_c = \frac{1}{2}h$ is called Nyquist frequency.

To eliminate the influence of masking (aliasing), there are two practical methods. One of them consists of selecting such a small magnitude h that the components with frequencies higher than f_c in the process under test are physically impossible. According to the second method the process is filtered before sampling. As a result, the filtered process will not contain components with frequencies higher than the highest frequency limit wanted. In this case, the frequency selection f_c equal to a maximum test frequency will yield the accurate results for frequencies below f_c. The second method is preferable to the first because it makes it possible to save processing time and decreases the cost of computations.

The interval of discreteness $h - \Delta t$ is selected so that

$$h = {}^1/_2 f_c$$

where $½f_c$ is the smallest period of realization. The selected h must be small enough to exclude the errors of masking. For accurate measurement of correlation function when this function contains frequencies close to f_c, the interval of discretization should be taken equal to $2.5f_c$. If the main emphasis is laid on the measurement of power spectrum then it is sufficient to take the interval of discreteness equal to $¼f_c$. It goes without saying that due to economic considerations one must strive to adopt the value being as close as possible to $½f_c$.

The maximum number of steps for correlation function is selected by the formula

$$m = 1/(B_e h)$$

where B_e is the equivalent of resolving power in computation of the power spectrum.

Thus for a constant value h, the magnitude B_e will decrease with increasing m. The volume of sampling N is such that

$$N = m/\varepsilon^2$$

where ϵ is the normalized mean square error, prescribed during the computation of the spectrum.

The respective minimum length of realization is $T_r = Nb$.

The number of degrees of freedom in computation of the estimates of the spectrum $n = 2N/m$.

The normalized mean square value of error is determined by the expression $\epsilon = \sqrt{m/N}$.

Under N terms of sequence $\{x_n\}$ $(n = 1, 2, \ldots, N)$ taken from the transformed realization $x(t)$, which is a stationary process and has the mean value $\bar{x} = 0$, the estimate of autocorrelation function at the shift rh is determined by the formula

$$\hat{R}_r = [1/(N - r)] \sum_{n=1}^{N-r} x_n x_{n+r}, \quad r - 0, 1, 2 \ldots, m$$

where r is the step; m is the maximum number of steps; R_r is the estimate of the true value of autocorrelation function R_r at the step r corresponding to the shift rh. It should be noted that if $x \neq 0$, and $x = (1/N) \Sigma_{n=1}^{N} x_n$, then the present sequence may be substituted by the sequence $x'_n = x_n - \bar{x}$; $n = 1, 2, 3, \ldots, N$.

The maximum number of steps m determines the equivalent resolution of spectral density within the frequency interval $(0, f_c)$. Thus the equivalent resolution B_e is equal to the double width of the interval, equal to quotient of frequency integral $(0, f_c)$ on ranges of integration separated from each other by f_c/m. Knowing the frequency beforehand, it is possible to select magnitude m so that the desired value B_e is obtained. In order to have the statistical uncertainty at the minimum during the forthcoming calculation of estimates of spectral density, it is necessary to select $m \ll N$ because the maximum number of statistical degrees of freedom is $2N/m$. On the other hand, in order to obtain high resolution (of small magnitude B_e), m should be chosen large. Hence, in choosing m a compromising solution has to be found. It is recommended that m should not be taken more than 1/10 of the volume of sample N. This makes it possible to prevent instability.

For the sample from the stationary transformed realization $x(t)$, with $\bar{x} \neq 0$ the initial estimate $\sigma_x(f)$ of the actual density $\sigma_x(f)$ is determined for arbitrary values f of the range $0 \leq f \leq f_c$ by the formula

$$\hat{\sigma}_x(f) = 2h\left[\hat{R}_0 + 2\sum_{r=1}^{m-1} \hat{R}_r \cos(\pi r f)/f_c + \hat{R}_m \cos(\pi f m)/f_c\right]$$

where h is the interval of time between the samplings; R_r is the estimate of autocorrelation function at the step r; m is the maximum number of steps.

It is recommended to compute the values of functions only for $m + 1$ of discrete frequencies

$$f = kf_c/m, \quad k = 0, 1, 2, \ldots, m$$

The final estimate may be found by the further smoothing of spectral density by frequency, for example, using the Hann method

$$\begin{gathered} \hat{\sigma}_0 = 0.5\hat{\sigma}_m + 0.5\hat{\sigma}_k \\ \hat{\sigma}_k = 0.25\hat{\sigma}_{k-1} + 0.5\hat{\sigma}_k + 0.25\hat{\sigma}_{k+1}, \quad k = 1, 2, 3, \ldots, m-1 \\ \hat{\sigma}_m = 0.5\hat{\sigma}_{m-1} + 0.5\hat{\sigma}_m \end{gathered}$$

where $\hat{\sigma}_k$ is the smoothed estimate for the harmonic of order k.

The other equivalent way of obtaining an estimate of applying Hann's weight function,

$$D_r = D(rh) = \begin{cases} 1/2\,[1 + \cos(\pi r/m)], & r = 0, 1, 2, \ldots, m \\ 0 & r > m \end{cases}$$

More practical from the point of view of using computers are formulas in which "weighing" is already taken into account while computing $\hat{\sigma}(f)$, i.e.,

$$\hat{\gamma}(f) = (1/\pi)\left[(1/2)\hat{R}_0 + (1/N)\sum^{m}(N-r)\hat{R}_r \right.$$
$$\left. \times (0.54 + 0.46\cos(\pi r/m)\cos fr\right]$$

The presented methods of control and diagnostics are based on the application of apparatus meant for the mathematical processing of the test data, since in recent years, along with the development of automatic design, the systems of automation for research work (diagnostics is part of it) have been developed.

Therefore, the processes taking place in bearing assemblies must be and can be described in terms of mathematical methods. This will enable us to solve any problem of control and diagnostics with a minimum amount of information derived from the test object.

REFERENCES

1. Artobolevskiy, I. I., Babrovinitskiy, Yu. I., and Genkin, M. D. Vvedeniye v akusticheskuyu dinamiku mashin (Introduction into acoustic dynamics of machines). Moscow, Nauka, 1979, 295 pages.
2. Beyzel'man, R. D., Tsypkin, B. V., and Perel', L. Ya. Podshipniki kacheniya (Rolling contact bearings). Moscow, Mashinostroyeniye, 1975, 574 pages.
3. Bowden, F. P. and Taybor, D. Treniye i smazka tvyordykh tel (The friction and lubrication of solids). Translated from the English, Moscow, Mashgiz, 1968, 543 pages.
4. Genkin, M. D. Voprosy akusticheskoy diagnostiki (On acoustic diagnostics). In: Metody izolyatsii mashin i prisoyedinitel'nykh konstruktsii (Isolation techniques for machines and attached structures). Moscow, Nauka, 1975, pp. 96–123.
5. Ragulskis, K. M., Yurkauskas, A. Yu., and Atstupenas, V. V. Vibratsii podshipnikov (Vibrations of bearings). Vilnius, Mintis, 1975, 392 pages.
6. Kapitsa, P. L. Gidrodinamicheskaya teoriya smazki pri kachenii (Hydrodynamic theory of lubrication in rolling). Zhurnal Tekhnicheskoy Fiziki, 1955, Vol. 25, No. 4, pp. 747–762.
7. Shenfel'd, A. Ya. and Yurkauskas, A. Yu. Graduirovka vysokotochnykh izmeriteley momentov (Calibration of high-precision torque meters). In: Nauchnye trudy vuzov Lit. SSR. Vibrotekhnika, 1977, No. 4(28), pp. 76–85.
8. Darguzhis, S. A. and Yurkauskas, A. Yu. Sredstva izmereniya parametrov podshipnikov (Means of measurement of bearing parameters). In deposited manuscript, Vilnius, Institute of Scientific and Technical Information, 1983, p. 90.
9. Evlanov, L. G. Priblizhennyy metod issledovaniya tochnosti lineynykh sistem, soderzhashchikh sluchaynye parametry (Approximate method of ascertaining the accuracy of linear systems incorporating random parameters). In: Optimal'nye sistemy. Statisticheskie metody. Moscow, Nauka, 1967, 114 pages.
10. Yurkauskas, A. Yu. Ragulskis, K. M., and Vitkute, A. E. Izmereniye i analiz momentov soprotivleniya i vibratsii pribornykh sharikopodshipnikov (Measurement and analysis of fric-

tion torque and vibration of instrument ball bearings). Nauchnye trudy vuzov Lit. SSR, Vibrotekhnika, 1970, No. 2(11), pp. 107–115.
11. Dovidenas, V. I. and Augutis, V. N. Izmereniye elektricheskovo soprotivleniya gidrodinamicheskoy maslyanoy plenki v podshipnikakh kacheniya (Measurement of the electrical resistance of a hydrodynamic oil film in rolling contact bearings). Nauchnye trudy vuzov Lit. SSR, Vibrotekhnika, 1969, No. 2(5), pp. 87–92.
12. Kakuta, K. Issledovaniye momenta treniya sharikopodshipnikakh, nakhodyashchikhsya pod osevoy nagruzkoy (Study of the friction torque of a ball bearing under an axial load). 1961, Vol. 27, No. 178, GPNTB No. 83390 (Translation), pp. 102–134.
13. Yurkauskas, A. Yu., Augutis, V. N., and Ziberkas, T. B. Issledovaniye razreshayushchiy sposobnosti izmereniya usiliyy pri pomoshchi tenzodatchikov (Study of the resolution of force measurement using strain gauges). Nauchnye trudy vuzov Lit. SSR, Vibrotekhnika, No. 4(17), 1972, pp. 363–368.
14. Yurkauskas, A. Yu. and Palnonis, A. Yu. Issledovaniye elektricheskovo soprotivleniya gidrodinamicheskoy plenki sharikopodshipnikov (Electrical resistance of the hydrodynamic film in ball bearings). Nauchnye trudy vuzov Lit. SSR, Vibrotekhnika, No. 4(12), 1970, pp. 15–20.
15. Kurushin, M. I. and Kodnir, D. S. Kinematika, napryazheniya i teplovydeleniye v radialno-upornykh sharikopodshipnikakh s uchetom vliyaniya smazki (Kinematics, stresses, and heat generation in radial-thrust ball bearings and the effect of lubrication). Trudy Kuibyshevskovo aviatsionovo in-ta, No. 40, 1969, pp. 53–64.
16. Kovalev, M. P. and Narodetskiy, M. Z. Raschet vysokotochnykh sharikopodshipnikov (Computation of high-precision ball bearings). Moscow, Mashinostroyeniye, 1980, 373 pages.
17. Kovalev, M. P., Sivokonenko, I. M., and Yavlenskiy, K. N. Opory priborov (Instrument bearings). Moscow, Mashinostroyeniye, 1968, 192 pages.
18. Kodnir, D. S. Kontaktno-gidrodinamicheskaya teoriya smazki (Contact-hydrodynamic theory of lubrication). Kuibyshev, Knizhnoye, 1963, 113 pages.
19. Kolebaniya i ustoychivost' uprugikh sistem mashin i priborov (Oscillations and stability of elastic systems of machines and instruments). Ed. M. V. Khvingiy. Tbilisi, Metsniyereba, 1974, 284 pages.
20. Lints, V. P. Tekhnicheskaya diagnostika mashin (Technical diagnostics of machines). Moscow, Znaniye, 1971, 46 pages.
21. Makarov, R. A. Sredstva tekhnicheskoy diagnostiki mashin (Means of technical diagnostics of machines). Moscow, Mashinostroyeniye, 1981, 183 pages.
22. Metod statisticheskikh ispitaniy (Metod Monte-Karlo) (Statistical test techniques. Monte-Carlo method). Ed. Yu. A. Shreider, Moscow, Fizmatgiz, 1962, 331 pages.
23. Raiko, M. V. and Trivailo, M. S. Metod izmereniya tolshchiny smazochnovo sloya v kontakte detaley mashin (Method for measurement of thickness of the lubricating layer on the contact surfaces of machine components). In: Fiziko-khimicheskaya mekhanika materialov, No. 5, 1965, pp. 19–24.
24. Reshchikov, B. F. Metodika issledovaniya koeffitsienta treniya kacheniya na rolikovoy mashine (Methodology for the study of rolling friction coefficients on a roller-type machine). In: Printsipy i novyye metody ispytaniy materialov dlya uzlov treniya. Moscow, Nauka, 1968, pp. 39–43.
25. Mozgalevskiy, A. V. and Gasarov, D. A. Tekhnicheskaya diagnostika (neprerybnyye obyekty) (Technical diagnostics/continuous objects). Moscow, Vysshaya shkola, 1975, 195 pages.
26. Rodinov, E. M. Momenty, voznikayushchie ot pogreshnosti formy poverkhnosti kacheniya podshipnikov (Torques due to defects in rolling surfaces of ball bearings). In: Trudy Moskovskovo aviatsionnovo tekhnologicheskovo in-ta, No. 59, Moscow, 1964, pp. 83–100.
27. Naryshkin, V. N., Starostin, V. F., and Grigor'yev, V. F. Radial'no-upornyye sharikopodshipniki dlya vysokotochnykh skorostnykh uzlov (Obzor) (Radial-thrust ball bearings for high-accuracy high-speed units (Review)). Moscow, Trudy NIIAvtoprom, 1979, 46 pages.
28. Kodnir, D. S. and Savvin, L. L. Opredeleniye kasatel'nykh nagruzok i koeffitsientov treniya

dlya poverkhnostey, rabotayushchikh v kontaktno-gidrodinamicheskom rezhime (Determination of tangential loads and friction coefficients for surfaces operating in a contact-hydrodynamic regime). In: Trudy Kuibyshevskovo aviatsionnovo in-ta. No. 40, 1969, pp. 62–69.

29. Yurkauskas, A. Yu. and Kranchyukas, R. B. Opredeleniye koeffitsientov zhestkosti poiskovym metodom (Determination of stiffness coefficients by the search method). Nauchnye trudy vuzov Lit. SSR. Vibrotekhnika, No. 4(17), 1972, pp. 195–198.
30. Spitsyn, N. A. and Mashnev, M. M. Opory osey i valov mashin i priborov (Bearings for axles and shafts of machines and instruments). Leningrad, Mashinostroyeniye, 1970, 519 pages.
31. Osnovy tekhnicheskoy diagnostiki (Fundamentals of technical diagnostics). Ed. P. P. Parkhomenko, Moscow, Energiya, Vol. 1, 1976, 460 pages.
32. Blekhman, I. I. and Mechasyan, S. A. O koeffitsientakh treniya pri vzaimodeystvii uprugovo tela s vibriruyushchey ploskost'yu (On friction coefficients during interaction of an elastic body with a vibrating plane). Mekhanika tverdovo tela, No. 4, 1970, pp. 60–83.
33. Pal'mgren, A. O nekotorykh svoystvakh podshipnikov kacheniya (Some characteristics of rolling contact bearings). Geteborg, 1961, translation GPNTB No. 28655, 46 pages.
34. Pavlov, B. V. Akusticheskaya diagnostika mekhanizmov (Acoustic diagnostics of mechanisms). Moscow, Mashinostroyeniye, 1971, 223 pages.
35. Pinegin, S. V. Opory kacheniya v mashinakh (Rolling contact bearings in machines). Moscow, Nauka, 1965, 263 pages.
36. Pinegin, S. V. Kontaktnaya prochnost' i soprotivlyaemost' kacheniyu (Contact strength and resistance to rolling). Moscow, Mashinostroyeniye, 1969, 243 pages.
37. Pribory i avtomaty dlya kontrolya podshipnikov (Instruments and automatic machines for bearings control). Moscow, Mashinostroyeniye, 1973, 256 pages.
38. Pribory dlya nerazrushayushchevo kontrolya materialov i izdeliy (Instruments for nondestructive evaluation of materials and products). Ed. V. V. Klyuyev. Moscow, Mashinostroyeniye, 1976, 326 pages.
39. Pribornyye sharikovyye podshipniki (Instrument ball bearings). Ed. K. N. Yavlenskiy, V. N. Naryshkin, A. A. Chayedayeva. Moscow, Mashinostroyeniye, 1981, 351 pages.
40. Prichiny shuma i vibratsiy v podshipnikakh kacheniya (Sources of noise and vibration in rolling contact bearings). Publication by offset duplicator of VNIIPP, 1967, 114 pages.
41. Yavlenskiy, A. K. Prognozirovaniye izmeneniya dinamicheskikh parametrov sharikopodshipnika (Forecast of dynamic parameter changes of a ball bearing). In: Raschety detaley priborov mekhanizmov. Mezhvuzovskiy sbornik, LIAP, No. 119, 1977, pp. 105–108.
42. Samokhin, O. N., Artyomkin, A. V., and Tolstov, A. E. Sharikopodshipniki malogabaritnykh skorostnykh elektodvigateliy (Ball bearings of high-speed electric motors). Moscow, NIIAvtoprom, 1977, 34 pages.
43. Serdakov, A. S. Avtomaticheskiy kontrol' i tekhnicheskaya diagnostika (Automatic monitoring and technical diagnostics). Kiev, Tekhnika, 1971, 146 pages.
44. Skorynin, Yu. A. Nadezhnost' i dolgovechnost' opor podvizhnykh sistem priborov (Reliability and longevity of instrument supports). Minsk, Nauka i tekhnika, 1965, 109 pages.
45. Sprishevskiy, A. I. Podshipniki kacheniya (Rolling contact bearings). Moscow, Mashinostroyeniye, 1969, 632 pages.
46. Augutis, V. N. and Yurkauskas, A. Yu. Sredstva i rezultaty eksperimental'nykh issledovaniy dinamiki rotornykh sistem (Means for and results of experimental study of dynamics of rotor systems). In: Trudy 6-ovo mezhdunarodnovo simpoziuma, Mishkol'ts, the Hungarian People's Republic, 1978, pp. 111–116.
47. Pinegin, S. V. and Frolov, K. V. Sovremennyye problemy vibratsii i shuma podshipnikov kacheniya (Contemporary problems of vibration and noise in rolling contact bearings). In: Vibratsii v mashinostroyenii i sel'skokhozyaystvennoy tekhnike. Erevan, 1966, pp. 43–68.
48. Treniye, iznashivaniye i smazka. Ed. I. V. Kragel'skii and V. V. Alisin. Moscow, Mashinostroyeniye, 1978 (Vol. 1), 1979 (Vol. 2), 358 pages.
49. Volkov, L. K. and Yavlenskii, K. N. Ustanovleniye korrelyatsii mezhdu vibratsionnymi i

akusticheskimi kharakteristikami sharikopodshipnikov (Determination of correlation between vibrating and acoustic characteristics of ball bearings). In: Trudy LIAP, 1976, pp. 144–148.

50. Yurkauskas, A. Yu. Metody i kriterii diagnoza podshipnikovykh uzlov (Diagnostic methods and criteria for bearing units). Deposited manuscript, Vilnius, Institute of Scientific and Technical Information, 1980, 88 pages.
51. Yurkauskas, A. Yu. Sredstva i metody diagnoza podshipnikovykh uzlov (Means and methods of diagnostics for bearing units). Deposited manuscript, Vilnius, Institute of Scientific and Technical Information, 1982, 188 pages.
52. Yavlenskiy, A. K. and Yavlenskiy, K. N. Teoriya dinamiki diagnostiki sistem trniya kacheniya (Theory of diagnostics dynamics for systems with rolling friction). Leningrad, Leningrad State University Press, 1978, 184 pages.
53. Bavden, F. P. and Taybor, D. The friction and lubrication of solids. Clarendon Press, Oxford, 1950, 231 pages.
54. Anderson, W. I. Elastohydrodynamic lubrication theory as a design parameter for rolling element bearings. ASME Paper, No. DE-19, 1970, pp. 97–108.
55. Palionis, A. and Yurkauskas, A. Erhohung der Funktionssicherarbeit von Wolzbagerungen.—Feingerätstechnik. GDR, Berlin, 1982, pp. 83–85.
56. Palmgren, A. Ball and roller bearing. Philadelphia, 1945, 194 pages.
57. Ragulskis, K. M., Yurkauskas, A., Atstupenas, V. V., Vitkute, A., and Kulves, A. P. Vibration of bearings. New Delhi, Oxanian Press pvt. ltd., 1979, 388 pages.

INDEX